Analog and Digital Filter Design Using C

Les Thede

Department of Electrical Engineering
Ohio Northern University

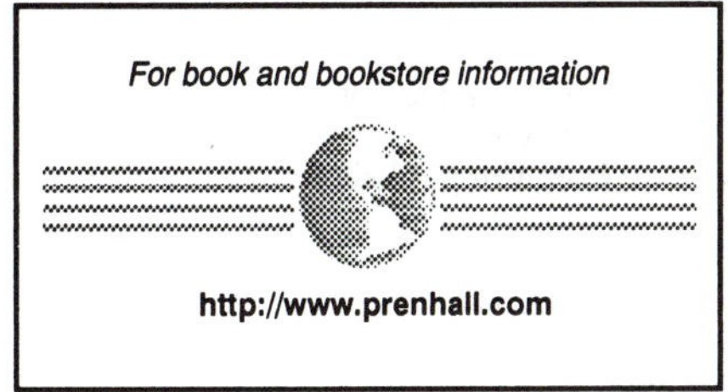
For book and bookstore information
http://www.prenhall.com

Prentice Hall PTR
Upper Saddle River, New Jersey 07458

Library of Congress Cataloging-in-Publication Data

Thede, Les (Leslie D.)
Analog and digital filter design using C / Les Thede.
p. cm.
Includes bibliographical references and index.
ISBN 0-13-352627-5
1. Electrical filters—Design and construction. 2. C (Computer program language) I. Title
TK7872.F5T53 1996
621.3815'324—dc20 95-24511
CIP

Editorial/production supervision: *Kerry Reardon*
Cover design director: *Jerry Votta*
Cover design: *Talar Agasyan*
Manufacturing buyer: *Alexis R. Heydt*
Acquisitions editor: *Karen Gettman*

Prentice-Hall, Inc.
A Simon & Schuster Company
Upper Saddle River, New Jersey 07458

The publisher offers discounts on this book when ordered in bulk quantities. For more information, contact:

Corporate Sales Department
Prentice Hall PTR
One Lake Street
Upper Saddle River, NJ 07458

Phone: 800-382-3419
Fax: 201-236-7141
E-mail: corpsales@prenhall.com

Printed in the United States of America

10 9 8 7 6 5 4 3 2 1

ISBN 0-13-352627-5

Prentice-Hall International (UK) Limited, *London*
Prentice-Hall of Australia Pty. Limited, *Sydney*
Prentice-Hall Canada Inc., *Toronto*
Prentice-Hall Hispanoamericana, S.A., *Mexico*
Prentice-Hall of India Private Limited, *New Delhi*
Prentice-Hall of Japan, Inc., *Tokyo*
Simon & Schuster Asia Pte. Ltd., *Singapore*
Editora Prentice-Hall do Brasil, Ltda., *Rio de Janeiro*

This text is dedicated to
LeRoy L. Thede,
the first of three generations
of electrical engineers.
Thanks, Dad!

Table of Contents

Preface

This book was intentionally written to be different from other filter design books in two important ways. First, the most common analog and digital filter design and implementation methods are ***covered in a no-nonsense manner***. All important derivations and descriptions are provided to allow the reader to apply them directly to his or her own filter design problem. Not only are the details of analog active and digital IIR and FIR filter design presented in an organized and direct manner, but implementation issues are discussed to alert the reader to potential pitfalls. The simulation of analog filters is made easier by the generation of PSpice circuit description files which includes R-C component values calculated directly from the filter coefficients. In addition, the testing of IIR and FIR filters designed for audio signals is enhanced by providing sample WAV and VOC sound files which can be filtered by using the digital filter design coefficients. Anyone with a sound card on their computer can then play the original and processed sound files for immediate evaluation.

The second difference between this book and others is that the important filter design and implementation techniques discussed in this text are ***supported by thoroughly tested C code***. No, the source code provided on the accompanying disk is not just a collection of fragmented functions, but rather a set of three organized programs to design and implement analog and digital filters. Not only are DOS executable programs provided on the disk, but also all of the source code is provided to allow additions or modifications to be made as desired. Virtually all of the included source code (except the graphics functions for frequency response display) is portable. Three graphics modules have been provided to support Microsoft, Borland, and MIX Software compilers on DOS platforms. However, readers using other platforms can still take advantage of the code in this text in three ways. They can forgo the frequency response display and still retain all of the design and implementation features of the software. Or they can link a graphics module for their system with the existing code for a complete system (requirements are presented in the text). And as a last option they can save the magnitude and phase responses to disk files (a standard feature) and display the data using other programs if desired. A basic knowledge of C programming is expected of the reader, but the code presented in the text is thoroughly discussed and well-documented. The text does assume the reader is familiar with the fundamental concepts of linear systems such as system transfer functions and frequency response although no prior knowledge of filter design is needed.

The construction of this text is unique in that the filter design and implementation techniques are developed by following the evolution of the C code for the three programs in the book. By following the development of

FILTER, the filter design program, throughout the text, the reader is introduced to the fundamental steps in analog and digital filter design. In addition, methods of frequency response determination and display are presented. The filter design problem is presented in ever-increasing detail until practically each entry in the project outline becomes a C function. The two filter implementation programs are developed in a similar manner. ANALOG aids in the implementation of analog active filters by determining electronic component values from the coefficients generated by FILTER and by generating PSpice circuit description files. DIGITAL illustrates the implementation of digital filters by allowing the user to digitally filter sound files using the coefficients generated by FILTER. The reader can then play the original and filtered sound files to hear the effects of the filter.

Chapter Contents

Chapter 1 introduces the reader to the filter design problem. An overview of the FILTER design project is presented with additional details of the project developed as they are needed in succeeding chapters. Chapter 2 develops the normalized transfer functions for the Butterworth, Chebyshev, inverse Chebyshev, and elliptic approximation cases. The FILTER project outline is enhanced to include the necessary C functions to determine the order and coefficients for these approximations. Chapter 3 describes the conversion of the normalized lowpass filter to an unnormalized lowpass, highpass, bandpass, or bandstop filter. By the end of the third chapter, a complete analog filter design can be performed.

Chapter 4 introduces the reader to the calculation of the frequency response of the analog filters designed in the previous chapters. In addition to the C code which calculates the frequency response, the reader is also introduced to the graphics functions necessary to display the frequency response. Techniques for compiling and linking the source files for the FILTER project are also discussed. In Chapter 5, the implementation of analog filters is considered using popular techniques in active filter design with discussion of real-world considerations. The ANALOG active filter implementation program is developed to determine the RC coefficients necessary to implement active filters. A PSpice circuit description file is generated to enable the filter developer to analyze the circuit. Chapter 5 completes the discussion of analog filters in this book.

Chapter 6 begins the discussion of discrete-time systems and digital filter design in this book. Several key features of discrete-time systems, including the notion of analog-to-digital conversion, Nyquist sampling theorem, the z-transform, and discrete-time system diagrams, are reviewed.

Similarities and differences between discrete-time and continuous-time systems are discussed. In Chapter 7, digital IIR (recursive) filters are designed. Three methods of designing IIR filters are considered with C code developed for the predominant bilinear transformation method. In addition, the frequency response calculations and related C code for the IIR filter are developed. Chapter 8 considers digital FIR (nonrecursive) filters using a variety of window methods and the Parks-McClellan optimization routine. The special techniques necessary for FIR frequency response calculation are discussed before developing the C code for the FIR filter design portion of the FILTER project. The implementation of real-time and nonreal-time digital FIR and IIR filters is discussed in Chapter 9. Implementation issues such as which type of digital filter to use, accuracy of quantized samples, fixed or floating point processing, and finite register length computation are discussed. Popular sound file formats are introduced and the C code necessary to process these sound files is generated. Users may then use the DIGITAL program to process sound files using the filter coefficients determined by FILTER. The reader can then hear the effects of filtering by replaying the original and processed sound files on a sound card.

ACKNOWLEDGMENTS

I would not have been able to complete this book without the help and support of a number of people. First, I thank the reviewers of this text who provided many helpful comments, both in the initial and final stages of development. These include Malcolm Slaney, Randy Crane, Paul Embree, John O'Donnell, Dave Retterer, and Dave Bogner.

I also thank the staff at Prentice Hall who have provided me with help and guidance throughout the publication process. These include Senior Editor Karen Gettman and her administrative assistant Barbara Alfieri.

I thank the faculty of the Department of Electrical Engineering at Ohio Northern University for their encouragement and support throughout this hectic and time-consuming process.

And, finally, I thank my wife Diane for all of her encouragement and support, and for the many hours of proofreading a text that made no sense to her!

TRADEMARKS

Chapter 1

Introduction to Filters and Filter Design Software

Everyone has probably come in contact with one type of filter or another in their lifetime. Maybe it was a coffee filter used to separate the grounds from the liquid, or perhaps an oil filter to remove contaminants from the oil of an engine. Anyone working in an office often filters the unimportant work from the important. In essence then the act of filtering is the act of separating desired items from undesired items. Of course when we discuss filters in this text, we are not talking about coffee, oil or paperwork, but rather electronic signals. The electronic filters we will be designing will separate the desirable signal frequencies from the undesirable.

There are many types of electronic filters and many ways that they can be classified. A filter's frequency selectivity is probably the most common method of classification. A filter can have a lowpass, highpass, bandpass, or bandstop response, where each name indicates how a band of frequencies is affected. For example, a lowpass filter would pass low frequencies with little attenuation (reduction in amplitude), while high frequencies would be significantly reduced. A bandstop filter would severely attenuate a middle band of frequencies while passing frequencies above and below the attenuated frequencies. Filter selectivity will be the focus of the first section in this chapter.

Filters can also be described by the method used to approximate the ideal filter. Some approximation methods emphasize low distortion in the passband of the filter while others stress the ability of the filter to attenuate the signals in the stopband. Each approximation method has visible characteristics which distinguish it from the others. Most notably, the absence or presence of ripple (variations) in the passband and stopband clearly set one approximation method apart from another. Filter approximation methods will be discussed in further detail in the second section.

Another means of classifying filters is by the implementation method used. Some filters will be built to filter analog signals using individual components mounted on circuit boards, while other filters might simply be part of a larger digital system which has other functions as well. Several implementation methods will be described in the third section of this chapter as well as the differences between analog and digital signals. However, it should be noted that digital filter implementation will be considered in detail starting in Chapter 6, while the first five chapters concentrate on filter approximation theory and analog filter implementation.

In the final sections of this chapter we will begin discussing the development of a C program to design analog and digital filters. In these sections, we will develop an approach to be used throughout the remainder of this text. A spiral technique will be employed to allow us to study the framework of the project in ever greater detail. This framework will not only include the primary functions to be employed within the program, but also the definition of the primary data variables used. By the end of this chapter we will have the framework of the main filter design program and the ability to enter and store the key filter specifications. We will even test FILTER, the filter design software included with this text.

1.1 Filter Selectivity

As indicated earlier, a filter's primary purpose is to differentiate between different bands of frequencies, and therefore frequency selectivity is the most common method of classifying filters. Names such as lowpass, highpass, bandpass, and bandstop are used to categorize filters, but it takes more than a name to completely describe a filter. In most cases a precise set of specifications is required in order to allow the proper design of a filter. There are two primary sets of specifications necessary to completely define a filter's response, and each of these can be provided in different ways.

The frequency specifications used to describe the passband(s) and stopband(s) could be provided in Hertz (Hz) or in radians/second (rad/sec). We will use the frequency variable f measured in Hertz as filter input and output specifications because it is a slightly more common way of discussing frequency. However, the frequency variable ω measured in radians/second will also be used as the internal variable of choice as well as for unnormalized frequency responses since most calculations will use radians/second.

The other major filter specifications are the gain characteristics of the passband(s) and stopband(s) of the filter response. A filter's gain is simply the ratio of the output signal level to the input signal level. If the filter's gain is greater than 1, then the output signal is larger than the input signal, while if the gain is less than 1, the output is smaller than the input. In most filter applications, the gain response in the stopband is very small. For this reason, the gain is typically converted to decibels (dB) as indicated in Equation 1.1. For example, a filter's passband gain response could be specified as 0.707 or –3.0103 dB, while the stopband gain might be specified as 0.0001 or –80.0 dB.

$$\text{gain}_{\text{dB}} = 20 \cdot \log(\text{gain}) \tag{1.1}$$

As we can see, the values in dB are more manageable for very small gains. Some filter designers prefer to use attenuation (or loss) values instead of gain values. Attenuation is simply the inverse of gain. For example, a filter with a gain of 1/2 at a particular frequency would have an attenuation of 2 at that frequency. If we express attenuation in dB we will find that it is simply the negative of the gain in dB as indicated in Equation 1.2. Gain values expressed in dB will be the standard quantities used as filter specifications, although the term attenuation (or loss) will be used occasionally when appropriate.

$$\text{attn}_{\text{dB}} = 20 \cdot \log(\text{gain}^{-1}) = -20 \cdot \log(\text{gain}) = -\text{gain}_{\text{dB}} \tag{1.2}$$

1.1.1 Lowpass Filters

Figure 1.1 shows a typical lowpass filter's response using frequency and gain specifications which are necessary for precision filter design. The frequency range of the filter specification has been divided into three areas. The passband extends from zero frequency (DC) to the passband edge frequency f_{pass}, and the stopband extends from the stopband edge frequency f_{stop} to infinity. (We will see later in this text, that digital filters have a finite upper frequency limit. We will discuss that issue at the appropriate time.) These two bands are separated by the transition band which extends from

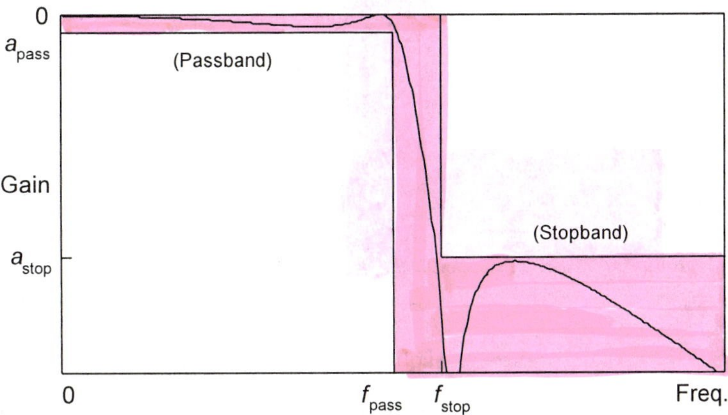

Figure 1.1 Lowpass filter specification.

f_{pass} to f_{stop}. The filter response within the passband is allowed to vary between 0 dB and the passband gain a_{pass}, while the gain in the stopband can vary between the stopband gain a_{stop} and negative infinity. (The 0 dB gain in the passband relates to a gain of 1.0, while the gain of negative infinity in the stopband relates to a gain of 0.0.) A lowpass filter's selectivity can now be specified with only four parameters: the passband gain a_{pass}, the stopband gain a_{stop}, the passband edge frequency f_{pass} and the stopband edge frequency f_{stop}.

Lowpass filters are used whenever it is important to limit the high frequency content of a signal. For example, if an old audio tape has a lot of high-frequency "hiss", a lowpass filter with a passband edge frequency of 8 kHz could be used to eliminate much of the hiss. Of course, it also eliminates high frequencies which were intended to be reproduced. We should remember that any filter can differentiate only between bands of frequencies, not between information and noise.

1.1.2 Highpass Filters

A highpass filter can be specified as shown in Figure 1.2. Note that in this case the passband extends from f_{pass} to infinity (for analog filters) and is located at a higher frequency than the stopband which extends from zero to f_{stop}. The transition band still separates the passband and stopband. The passband gain is still specified as a_{pass} and the stopband gain is still specified as a_{stop}.

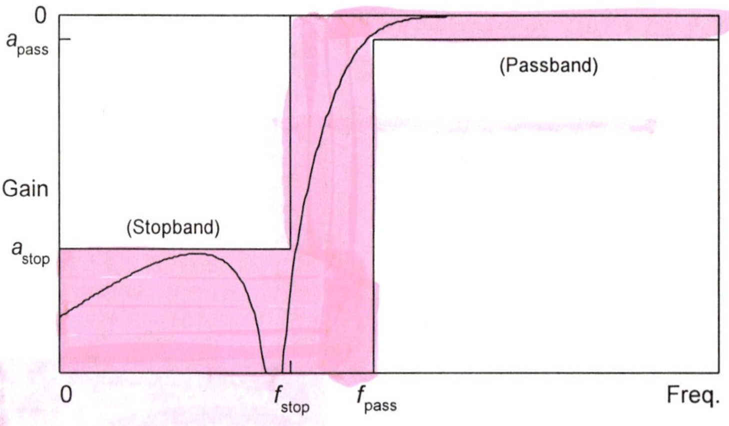

Figure 1.2 Highpass filter specification.

Highpass filters find a use when it is important to eliminate low frequencies from a signal. For example, when turntables are used to play LP records (some readers may remember those black vinyl disks which would warp in a car's back window), turntable rumble can sometimes occur, producing distracting low-frequency signals. A highpass filter set to a passband edge frequency of 100 Hz could help to eliminate this distracting signal.

1.1.3 Bandpass Filters

The filter specification for a bandpass filter shown in Figure 1.3 requires a bit more description. A bandpass filter will pass a band of frequencies while attenuating frequencies which are above or below that band. In this case the passband exists between the lower passband edge frequency f_{pass1} and the upper passband edge frequency f_{pass2}. A bandpass filter has two stopbands. The lower stopband extends from zero to f_{stop1}, while the upper stopband extends from f_{stop2} to infinity (for analog filters). Within the passband, there is a single passband gain parameter a_{pass}. However, individual parameters for the lower stopband gain a_{stop1} and the upper stopband gain a_{stop2} could be used if necessary.

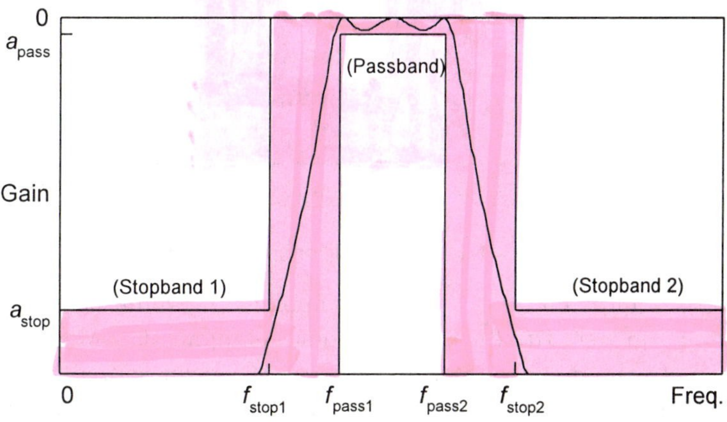

Figure 1.3 Bandpass filter specification.

A good example for the application of a bandpass filter is the processing of voice signals. The normal human voice has a frequency content located primarily in the range of 300 – 3000 Hz. Therefore, the frequency response for any system designed to pass primarily voice signals should contain the input signal to that frequency range. In this case, f_{pass1} would be 300 Hz and f_{pass2} would be 3000 Hz. The stopband edge frequencies would be selected by how fast we would want the signal response to roll off above and below the passband.

1.1.4 Bandstop Filters

The final type of filter to be discussed in this section is the bandstop filter as shown in Figure 1.4. In this case the band of frequencies being rejected is located between the two passbands. The stopband exists between the lower stopband edge frequency f_{stop1} and the upper stopband edge frequency f_{stop2}. The bandstop filter has two passbands. The lower passband extends from zero to f_{pass1}, while the upper passband extends from f_{pass2} to infinity (for analog filters). Within the stopband, the single stopband gain parameter a_{stop} is used. However, individual gain parameters for the lower and upper passbands, a_{pass1} and a_{pass2}, respectively, could be used if necessary.

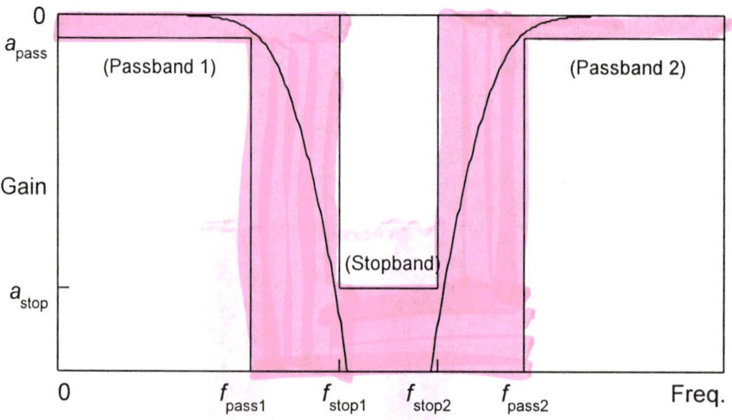

Figure 1.4 Bandstop filter specification.

An excellent example of a bandstop application would be a 60 Hz notch filter used in sensitive measurement equipment. Most electronic measurement equipment today runs from an AC power source using a 60 Hz input frequency. However, it is not uncommon for some of the 60 Hz signal to make its way into the sensitive measurement areas of the equipment. In order to eliminate this troublesome frequency, a bandstop filter (sometimes called a notch filter in these applications) could be used with f_{stop1} set to 58 Hz and f_{stop2} set to 62 Hz. The passband edge frequencies could be adjusted based on the other technical requirements of the filter.

1.2 Filter Approximation

The response of an ideal lowpass filter is shown in Figure 1.5, where all frequencies from 0 to f_o are passed with a gain of 1, and all frequencies above

f_0 are completely attenuated (gain = 0). This type of filter response is physically unattainable. Practical filter responses which can be attained are also shown. As a filter's response becomes closer and closer to the ideal, the cost of the filter (time delay, number of elements, dollars, etc.) will increase. These practical responses are referred to as approximations to the ideal. There are a variety of ways to approximate an ideal response based on different criteria. For example, some designs may emphasize the need for minimum distortion of the signals in the passband and would be willing to trade-off stopband attenuation for that feature. Other designs may need the fastest transition from passband to stopband and will allow more distortion in the passband to accomplish that aim. It is this engineering trade-off which makes the design of filters so interesting.

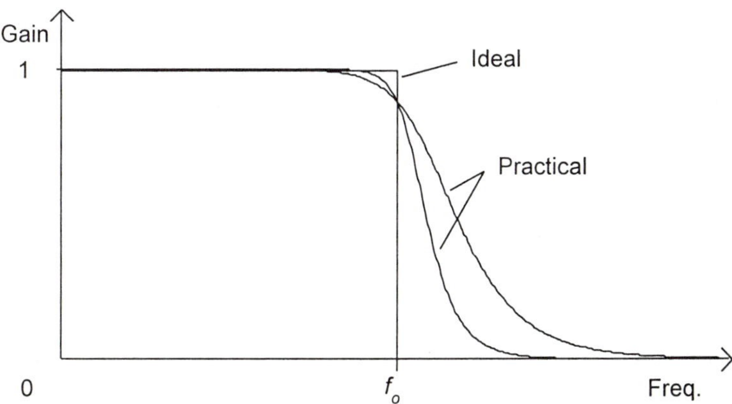

Figure 1.5 Practical and ideal filter responses.

We will be discussing the primary approximation functions used in filter design today which can be classified both by name and the presence of ripple or variation in the signal bands. Elliptic or Cauer filter approximations provide the fastest transition between passband and stopband of any which are studied in this text. An illustration of the magnitude response of an elliptic filter is shown in Figure 1.1, where we can see that ripple exists in both the passband and stopband. What is not shown in that figure is the phase distortion that the elliptic filter generates. If the filter is to be used with audio signals, this phase distortion must usually be corrected. However, in other applications, for example the transmission of data, the elliptic filter is a popular choice because of its excellent selectivity characteristics. The elliptic approximation is also one of the more complicated to develop. (We will discuss all approximation methods in greater detail in Chapter 2.)

The inverse Chebyshev response is another popular approximation method which has a smooth response in the passband, but variations in the stopband. On the other hand, the normal Chebyshev response has ripple in the passband, but a smooth, ever-decreasing gain in the stopband. The phase distortion produced by these filters is not as severe as for the elliptic filter, and they are typically easier to design. The inverse Chebyshev response is shown in Figure 1.2, and the normal Chebyshev response is illustrated in Figure 1.3. The Chebyshev approximations provide a good compromise between the elliptic and Butterworth approximation.

The Butterworth filter is a classic filter approximation which has a smooth response in both the passband and stopband as shown in Figure 1.4. It provides the most linear phase response of any approximation technique discussed in this text. (The Bessel approximation provides better phase characteristics, but has very poor transition band characteristics.) However, as we will see in Chapter 2, a Butterworth filter will require a much higher-order to match the transition band characteristics of a Chebyshev or elliptic filter.

1.3 Filter Implementation

After a filter has been completely specified, various numerical coefficients can be calculated (as described in Chapter 2). But after all the paperwork has been completed, the filter still has to be placed into operation. The first major decision is whether to use analog or digital technology to implement the filter. The differences between analog and digital filter design are based primarily on the differences between analog and digital signals themselves. Any signal can be represented in the time domain by plotting its amplitude versus time. However, the amplitude and time variations can be either continuous or discrete. If both the amplitude and time variations are continuous as shown in Figure 1.6 (a), the signal is referred to as an analog signal. Most real-life signals are analog in nature, for example, sounds that we hear, electrocardiogram signals recorded in a medical lab, and seismic variations recorded on monitoring equipment. However, the problem with analog signals is that they contain so much information. The exact amplitude of the signal (with infinite precision) is available at every instant of time. Do we actually need all of that information? And how do we store and transfer that information?

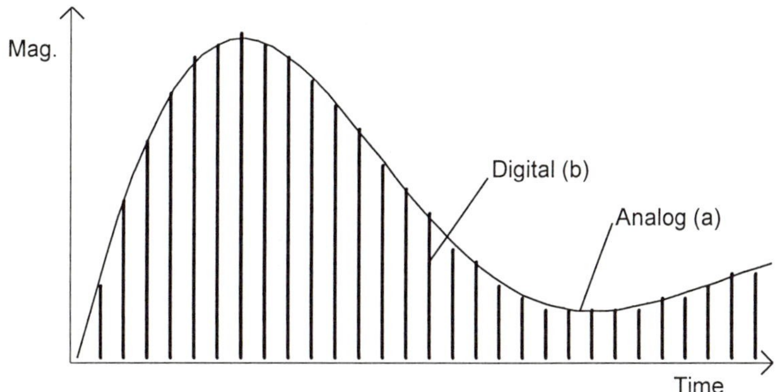

Figure 1.6 Comparison of (a) analog and (b) digital signals.

As techniques for the storage and transmission of information in digital form are becoming more efficient and cost effective, it is increasingly advantageous to use signals which are in a digital form. The advantage of using the resulting digital signals is that the amount of information can be managed to a level appropriate for each application. An analog signal can be converted to a digital signal in two steps. First, the signal must be sampled at fixed time intervals, and then the amplitude of the signal must be quantized to one of a set of fixed levels. Once the analog signal has been converted to a discrete-time and discrete-amplitude signal it is commonly referred to as a digital signal as shown in Figure 1.6 (b). The operation of sampling and quantizing is accomplished by an analog-to-digital converter (ADC). After filtering the signal in the digital domain, a digital-to-analog converter (DAC) can be used to return the signal to analog form. (A more complete discussion of these operations will be given in Chapter 6 where digital filter design is introduced and in Chapter 9 where practical considerations of digital filter implementation are discussed.) Today the digital images and sound files on computers as well as the music on compact disks are examples of signals in digital form.

If we choose to implement a filter in analog form, we still have further choices to make. We could choose to implement the filter with purely passive components such as resistors, capacitors and inductors. This approach might be the best choice when high frequencies or high power is used. In other circumstances, analog active filters might be the best choice where either transistors or operational amplifiers are used to provide a gain element in the filter. The implementation of analog active filters is considered in Chapter 5.

Digital filters will be implemented by using digital technology available today. Generally, the process will take place within a microprocessor system which could have other functions besides the filtering of signals. We will discuss two basic types of digital filters in Chapters 7 and 8 of this text. The first is the infinite impulse response (IIR) digital filter which is based in a large part on the design methodology of analog filters. The second type is the finite impulse response (FIR) digital filter which uses a completely different method for its design. The implementation of digital filters is considered in Chapter 9.

As we can now see, there is more to describing a filter than referring to it as a lowpass filter. For example, we might be designing an “analog active lowpass Butterworth filter” or a “digital IIR bandpass Chebyshev filter.” These names along with a filter's specification parameters will completely describe a filter.

1.4 FILTER DESIGN USING C

Although learning analog and digital filter design techniques is the first objective of this text, we are now ready to learn more about the second. We will see in the succeeding chapters that the design of analog or digital filters can be a tedious task and automated methods are required. Filter design software is available from a number of sources, but it can be expensive and is not always exactly what we need. Therefore, this text will not only discuss how both analog and digital filters are developed, but also how to write C functions to carry out the complex mathematical calculations necessary. In addition to the calculation of the analog and digital filter coefficients, C code will also be developed to view the frequency and phase responses of the filters and to implement the analog and digital filters.

The C programming language has become a predominant force in engineering and computer science in the past 5 – 10 years. In particular, it is the primary language used in digital filter design (with the possible exception of hardware-specific assembly language). C provides the combination of higher-level constructs while producing fast, compact executables. Other languages are available, and could be used in filter design, but C provides the best combination of efficiency and effectiveness. This text is written with the assumption that readers have a basic understanding of programs written in C. Coding practices which are more advanced will be discussed as appropriate. Several C language references are listed in Appendix A.

If we want our filter design software to be broad based in its capabilities, technically accurate, expandable, and visually informative, we must have a well-defined project. The approach which will be used in this text is a spiral one which works well for large projects. In this technique, the entire project will be divided into major functional blocks. Then as we continue our development, each of these major blocks will be subdivided into smaller and smaller components. This procedure will continue until we reach the point where individual C functions are written to accomplish a specific task. In addition to specifying the functional requirements at the outset of the project, we must also specify the major data elements which will be used to carry the filter information from one functional block to the next.

1.4.1 Project Data Definition

The primary data elements describing our filter will be used by a number of the design functions and therefore should be made available in a neat package. An array would be nice, but arrays require that all elements are of the same data type. Therefore the only reasonable choice is a structure. One method of defining the structure in C is as follows:

```
struct Sample
{ int      int_item1,int_item2;
  float    flt_item1,flt_item2;
} ;
```

Here the structure is defined with the `struct` keyword and includes member variables of different types. Since not all structures are alike (in fact we can define them in any way we want), we include the tag `Sample` to indicate that this is a particular type of structure. This would work fine except that everywhere we used this structure (and there will be numerous occasions in our project) we will have to use the cumbersome notation -- `struct Sample` -- instead of just the name `Sample`. However, we can use the `typedef` statement to allow us to give different names to data types. For example, if we become tired of defining variables `x`, `y`, and `z` as `unsigned long int`'s, we can use a `typedef` statement to define that data type as a `ul` and thereafter use the much shorter abbreviation as shown below.

```
typedef unsigned long int ul;
ul x, y, z;
```

In a similar manner we can define our structure as a `Sample` data type and remove the need for the additional word `struct` with all of our declarations. The new definition of the `Sample` structure, as well as variables of that type, are shown below. Note that `A` is a structure of the type `Sample`, while `B` is a pointer to a structure of the type `Sample`.

```
typedef struct
{ int   int_item1,int_item2;
  float flt_item1,flt_item2;
} Sample;

Sample A,*B;
```

We will store our filter data elements in a structure called `Filt_Params`. We can assume that we will need all of the filter specifications as outlined earlier in this chapter. These will include the passband and stopband edge frequencies (`wpass1, wpass2, wstop1,` and `wstop2`) and gains (`apass1, apass2, astop1,` and `astop2`). We will also need indicators of the filter selectivity, the approximation method, and the implementation type (`select, approx,` and `implem`). The sampling frequency (`fsamp`) is an additional element which will be necessary for digital filters, but will not be used by analog filters. All of these variables represent the input specification for the filter design and are shown in the `Filt_Params` structure below. The data type `double`, which provides 15 significant digits, is used for all floating point variables and should provide the necessary accuracy for our filter design. The filter's type indicators will be stored as `char`'s.

```
typedef struct
{ double  apass1,apass2,    /*  passband gain's */
          astop1,astop2,    /*  stopband gain's */
          wpass1,wpass2,    /*  passband edge freq's */
          wstop1,wstop2,    /*  stopband edge freq's */
          fsamp,          /*  samp freq for dig filt */
          gain,           /*  gain multiplier */
          *acoefs,*bcoefs;  /*  ptr's to coefs */
  int     order;     /*  order or length of filter */
  /* selectivity, approximation and implementation */
  char    select,approx,implem;
} Filt_Params;
```

In addition to the input specifications, there will also be a need for another set of filter data once the filter has been designed. The filter's order (`order`) indicates the size of the filter's transfer function and is stored as an integer. The filter's gain constant (`gain`) and pointers to two sets of filter coefficients (`acoefs` and `bcoefs`) will completely describe the filter's transfer function and will be stored in the structure as `double`'s. (Perhaps some of these data items may not be clear at this point, but rest assured that they will be completely described in the chapters to come.)

It is impossible to determine the number of coefficients necessary to describe a filter before the filter is designed. We will design some filters with fewer than 10 coefficients, but other designs may require more than 200 coefficients. For that reason we will use pointers to the arrays instead of simply defining `acoefs` and `bcoefs` as large arrays. Using large arrays would make the structure unnecessarily large, and there would still be no guarantee that these fixed arrays would be large enough to store every filter which might be designed. It will be far more efficient to determine the size of the array necessary for the individual design, then allocate memory dynamically for the coefficient array and finally store the pointer (address) to the array in the structure. That way only a pointer variable needs to be stored in the structure, not an entire array.

As it is, our `Filt_Params` structure is fairly large, and since we will be using this structure with most of the functions which we develop, it would be much more efficient to transfer a pointer (address) to the structure instead of the structure itself. Actually, there is another reason for transferring a pointer to the structure rather than the structure itself. In C all arguments of functions are "called by value," which means that a copy of the argument is transferred to the function. This practice does not allow the transferred variable to be changed by the function. (Technically, we can change the variable all we want within the function, but when we leave the function the original value of the variable within the calling function remains unchanged.) Since our filter design functions will need to make changes to the data values stored in the structure, sending a copy of the structure to the function without allowing data values to change would be unacceptable. However, if we send a pointer to the structure (or actually a copy of the address), we will be able to access the actual memory locations of the data stored in the structure. In this way, our filter design functions will be able to enter values into the structure as necessary.

The use of a structure in this manner allows our filter design functions to receive input and provide output via the structure. Otherwise we would find it difficult to have our functions return several values at the same time. Another method would require two structures instead of one. This would allow the separation of data elements into an input and output set and therefore limit the amount of unused information provided to the filter design functions. In fact, the initial development of this project was following that method. However, when the list of design functions was reviewed, most of them used both structures as arguments, and some parameters were duplicated in both structures. Therefore, since little efficiency was gained by using two structures, the simpler and clearer method of using only one structure was selected.

1.4.2 Project Procedure Overview

Sometimes it is difficult to look at the global overview of a project, especially if our attention has been focused on the intricate details. But the attempt in this section is to pull back far enough to see the entire project. Then we will spiral in on the project, providing ever-increasing detail until all functions for the project have been written. As a start, consider the following outline of our filter design project.

Analog and Digital Filter Design Project

I. Design analog or digital filter.
II. Determine frequency response of filter.
III. Display magnitude and phase response on screen.

This initial outline serves the purpose of focusing our attention on the global task at hand, but it does not provide the necessary detail which the project definition requires. Each of these major outline headings needs to be broken into subheadings for further clarification. For example, let's expand the first item to the next level as shown in the outline below.

Analog and Digital Filter Design Project

I. Design analog or digital filter.
    A. Get filter specifications from user.
    B. Design filter according to specifications.
    C. Display filter parameters on screen.
    D. Save filter parameters to file.
II. Determine frequency response of filter.
III. Display magnitude and phase response on screen.

This level of specification is still too abstract, so let's continue the spiraling technique and expand each of the four subdivisions even further. At this point we all have a good idea of what will be required of this filter design project, at least in the design portion of the program. Although more details will be necessary, we can develop those in a later chapter.

Analog and Digital Filter Design Project

I. Design analog or digital filter.
  A. Get filter specifications from user.
    1. Get implementation type (analog/dig IIR/dig FIR).
    2. Get selectivity type (LP/HP/BP/BR).
    3. Get approximation type(Buttr/Chby/IChby/Ellip).
    4. Get passband and stopband edge frequencies.
    5. Get passband and stopband gains.
    6. Get a descriptive title for project.
  B. Design filter according to specifications.
    1. Design analog, if designated.
    2. Design digital FIR, if designated.
    3. Design digital IIR, if designated.
  C. Display filter parameters on screen.
    1. Display filter input specifications.
    2. Display filter design values.
  D. Save filter parameters to file.
    1. Save parameters in text form.
    2. Save parameters in binary form.
II. Determine frequency response of filter.
III. Display magnitude and phase response on screen.

As the outline indicates, we must first determine what type of filter is to be designed and what specifications the filter is to meet. This will require us to ask the user for information, verify the information, and finally store the information in the `Filt_Params` structure. Once the filter specifications have been obtained from the user, the next step is to actually design the required filter. This step will need further division to be useful. Decisions must be made based on the implementation of filter, the selectivity type and the approximation type. Once the filter has been designed, the complete set of filter parameters must be presented to the user on the screen. The screen could then be printed for a hard copy, or the filter parameters may need to be saved to a data file as indicated in the fourth step. The saved parameters could take on one of two possible file formats (or both). The file format could just be a duplicate of the screen presentation which is easy to read, or it could be put into binary form. The binary form could be used to easily transfer the information into other programs (for example, digital hardware assembly language programs or circuit analysis programs).

1.5 C Code for FILTER Project

All of the code which we discuss is included on the computer disk supplied with this text. (Please read the x_README.TXT files in each directory for any important information.) The computer disk has subdirectories for each of the projects discussed in this text. The \FILTER directory includes all of the code for the filter design project, while the \ANALOG and \DIGITAL directories include the code for analog and digital filter implementation. Digitized sound files for use with the digital filter implementation can be found in the \SOUND directory. In addition, executable files (FILTER.EXE, ANALOG.EXE and DIGITAL.EXE) are located in the root directory of the disk. As we progress through these software projects, some listings of the project code will be shown, but because of space considerations, others will not. If the code is not included in the text, a description of where to find it on the disk will be given.

Now let's get started and build up our filter design program. We will generate numerous C functions before we reach the end of this text; so in order to keep this project manageable, it will be necessary to organize those functions into a number of modules. Each module will contain those functions which have a common purpose. Listed below are the names of the modules used throughout this text. All of these modules have their own header files with the same name except for an extension of .H. In addition to those header files, ERRORNUM.H is used to hold all of the error numbers used by the functions in the project.

ADV_MATH.C — contains advanced math functions.
ANALOG.C — contains analog filter implementation functions.
B_GRAPHX.C — contains graphics functions for Borland compiler.
COMPLEX.C — contains complex number functions.
DIGITAL.C — contains digital filter implementation functions.
FILTER.C — contains the filter design main function.
F_DESIGN.C — contains filter design functions.
F_RESPON.C — contains frequency response functions.
F_SCREEN.C — contains screen display functions.
GET_INFO.C — contains functions for acquiring information.
M_GRAPHX.C — contains graphics functions for Microsoft compiler.
UTILITY.C — contains utility functions.
X_GRAPHX.C — contains graphics functions for MIX compiler.

The source code modules (with extension of .C) contain the function definitions with brief descriptions of the functions and internal comments of the code. The header modules (with extension of .H) contain a list of other include files needed by the source file, a list of specially defined variables, structure definitions, and function prototypes for the functions in the related source files.

1.5.1 FILTER main Function

As indicated earlier, our initial code will be a framework which will evolve into the final program as we develop more and more details. Let's look at the FILTER `main` function which has been edited to include only the most important parts of the filter design project.

```
/*=====================================================
  Filename: FILTER.C
  Include:  FILTER.H
  Descript: Module contains the main() function for
            designing analog and digital (FIR & IIR)
            filters from user supplied filter params.
=====================================================*/
#include "filter.h"

/*=====================================================
  MAIN()
=====================================================*/
void main(void)
{ char   ans,          /* stores answer to question */
        *basename,     /* filename without extension */
        *descript;     /* title for output */
  int    Error;        /* error number */
  double Gain_Min;     /* min gain to be plotted */
  Filt_Params *FP;     /* ptr to filt params struct */
  Resp_Params *RP;     /* ptr to resp params struct */
  Scrn_Params *SP;     /* ptr to scrn params struct */

/* Alloc memory for the filter params struct */
  FP = (Filt_Params *)calloc(1,sizeof(Filt_Params));
  if(!FP)
  { Bail_Out(
      "Unable to allocate memory for FP!",ERR_ALLOC);
  }
  /* Get the filter specs from user, store in FP */
  Error = Get_Filter_Specs(FP);
  if(Error)
  { Bail_Out("Error in Get_Filter_Specs!",Error);}
```

```
    /*  Get title, store in descript */
    descript = Get_Fixed_String(
          "\n Please enter a title (40 char max):",40);
    if(!descript)
    { Bail_Out("Error in Get_Title!",ERR_NULL);}
    /*  Calc filter coefs, store in FP */
    Error = Calc_Filter_Coefs(FP);
    if(Error)
    { Bail_Out("Error in Calc_Filter_Coefs!",Error);}
    /*  Display both params and data to screen */
    Display_Filter_Params(FP,descript);
    /*  If user wants to save data, get filename */
    /*  If filename is zero-length, set back to NULL*/
    basename = Get_FILTER_Basename();
    if(!basename)
    { Bail_Out(
          "Error in Get_FILTER_Basename!",ERR_NULL);}
    if(strlen(basename) == 0)
    { free(basename);
      basename = NULL;
    }
    /*  Save filter data if filename specified */
    Error = Save_Filter_Params(FP,descript,basename);
    if(Error)
    { Bail_Out("Error in Save_Filter_Params!",Error);
    }
    Bail_Out("",0);
}
```

Listing 1.1 Segment of FILTER main function.

Before any of the primary functions in FILTER `main` are called, memory must be allocated for the `Filt_Params` structure used in the program. If for some reason no memory can be allocated and the `calloc` function returns a NULL pointer, the program execution must be scrubbed. This is accomplished by using the `Bail_Out` function which is part of the UTILITY.C module. The memory allocation is handled by `calloc` instead of `malloc` so that all data items within the structure are set to zero including the pointers. This provides a predefined state for the data items which is safer if one of the items is accessed before it is initialized. In most cases, zero will produce some undesirable result which will be caught by the programmer. If all goes well, `FP` will hold the address in memory where the structure is located. It is then sufficient to pass that one variable `FP` as an argument for functions which will then have access to all members of the structure, including the ability to change their values.

The Get_Filter_Specs function which is part of the F_DESIGN.C module is called next to obtain user information which is placed into the Filt_Params structure. Most of the functions used in this program will return an integer value of zero if no errors have occurred in the function. Any return value is stored in Error, and if it is zero, the program continues its normal execution path. If an error has occurred, the Bail_Out function will exit the program with an error message and number.

After getting the user's filter specifications, Get_Fixed_String is used to get a title (called descript) to use on all screen and file output. If a NULL string is returned (indicating an error), Bail_Out will alert the user. The Get_Fixed_String function is one of several functions in the GET_INFO.C module which have been written to obtain input from the user in a more structured manner than with the input functions in the standard C library.

The filter coefficients are calculated next and placed in the Filt_Params structure using the Calc_Filter_Coefs function which returns a zero value if no errors are encountered. The input specifications and calculation results can now be displayed by Display_Filter_Params as well. More discussion of these functions which are part of the F_DESIGN.C module will be postponed until the next chapter.

In many cases the user would like to save the filter parameters determined by the filter design. However, before the parameters can be saved, a filename must be obtained from the user. This has been made optional because in some cases a user might not want to save all filter design attempts. Therefore if the user does not want to save the program's results, the user simply hits the "Enter" key when asked for a filename, otherwise, a filename is entered. If the Get_FILTER_Basename returns a zero-length string for the filename, it is assumed that the user did not want to save the results and the basename pointer is reset to NULL. We will discuss the Get_FILTER_Basename which is part of the FILTER.C module in Chapter 3.

1.5.2 Get_Filter_Specs Function

The Get_Filter_Specs function will not be listed here because of its length, but the code can be found in \FILTER\F_DESIGN.C. (The Get_Filter_Specs function can be viewed using a standard text editor or a word processor.) The length of the function is due to all of the possible options available as the user specifies the type of filter desired. In addition, all user input is checked as it is entered to verify its validity. The basic outline for this function is given below. However, the descriptive title referred to in item I.A.6 is accepted in the `main` function.

I. Design analog or digital filter
   A. Get filter specifications from user.
      1. Get implementation type (analog/dig IIR/dig FIR).
      2. Get selectivity type (LP/HP/BP/BR).
      3. Get approximation type(Buttr/Chby/IChby/Ellip).
      4. Get passband and stopband edge frequencies.
      5. Get passband and stopband gains.
      6. Get a descriptive title for project.

A few comments are in order before we move on to other parts of the filter design program. In order to make this function as neat and clean as possible, we will use basic functions for getting user data such as numbers and letters which will be the primary responses solicited. These functions are all contained in the \FILTER\GET_INFO.C module. For example, Listing 1.2 shows a portion of the `Get_Filter_Specs` function which accepts the user's filter implementation choice. Notice that we can get access to the `implem` member of the `Filt_Params` structure by using the notation

```
/*  Ask user for implementation,  convert to
  uppercase. if OK, break; else, beep & try again*/
  FP->implem = ' ';
  while(FP->implem == ' ')
  { printf("\n Please select filter implementation:");
    printf(
        "\n\tAnalog (A)  Dig FIR (F)  Dig IIR (I)");
    FP->implem =Get_Char("\n Filter implementation: ");
    FP->implem = (char)toupper((int)FP->implem);
    switch (FP->implem)
    { case 'A':
      case 'F':
      case 'I': break;
      default:  FP->implem = ' ';
                printf("\a");
    }
  }
```

Listing 1.2 Segment of Get_Filter_Specs function.

FP->implem. This notation employs FP to locate the structure in memory, and ->implem finds the specific member of the structure. FP->implem is initially set to a blank character, and then we enter a while loop until the user provides a legitimate selection. After printing some instructions, we use the Get_Char function to obtain the user's response and convert it to an uppercase letter. The switch statement then checks for a legitimate response and we leave the switch with a legitimate value or with FP->implem set back to a blank. Leaving the switch statement with a blank will continue the while loop, but exiting the switch with an 'A', 'F' or 'I', will allow us to leave the while loop. Similar techniques are used throughout the remainder of the function to solicit and verify user input.

There are a number of global variables defined within the F_DESIGN.H file which are used in the Get_Filter_Specs function, so let's take a look at them now.

```
/*  #DEFINES:                    */

#define PI   3.1415926535898 /*  pi */
#define PI2 6.2831853071796 /*  2 * pi */
#define IOTA       1.0E-3   /* small increment */
#define FREQ_MIN  1.0E-3   /* min freq */
#define FREQ_MAX  1.0E12   /* max freq */
#define GAIN_PASS -1.0E-2  /* min passband gain (dB)*/
#define GAIN_TRAN -3.0103  /* min sb,max pb gain (dB)*/
#define GAIN_STOP -2.0E02  /* max stopband gain (dB)*/
```

Listing 1.3 #DEFINES section of F_DESIGN.H.

The #define statement is a C preprocessor directive which allows for the definition of important constants in a program. At compile time, the compiler will find every occurrence of PI and replace them with 3.1415926535898. The constant PI2 is defined as well as IOTA which is used to designate a very small increment. The minimum and maximum acceptable frequencies and limits for gain are also defined. A legitimate passband gain must be between GAIN_PASS and GAIN_TRAN, while a valid stopband gain must be between GAIN_TRAN and GAIN_STOP. These frequency and gain definitions help to solicit valid information from the user. If the user specifies a value outside of these established limits, chances are very good that it is an error in typing. It is better to check the information as it is supplied than to use invalid information in the filter design. Of course, these limits can be changed if desired for a specific filter application. This is the reason these variables are defined using the #define technique. By using this method, the variables need only be changed in one location, rather than searching the entire program for every occurrence.

1.5.3 Other Support Functions

There are a number of functions used in the filter design project which solicit input from the user. Let's take a look at the `Get_Double` function shown in Listing 1.4 now. (All of the similar functions used in this project are contained in the \FILTER\GET_INFO.C module.) In this function, we enter the infinite `while` loop, print the prompt `string`, and flush the input

```
/*=========================================================
  Get_Double() - prompts user for number, gets a
  numeric value which falls between min and max
  Prototype:  double Get_Double(char *string,
                                 double min,double max);
  Return:     the accepted double from keyboard
  Arguments:  string - a prompt for the user
              min - minimum acceptable value for input
              max - maximum acceptable value for input
=========================================================*/
double Get_Double(char *string,double min,double max)
{ int n;        /*  number of variables read by scanf */
  double dvalue;  /*  value to hold input */

  /*  prompt user, flush keyboard, scanf() must
  return 1 value between min and max. if OK, break
  from loop; if not, beep and try again */
  while(1)
  { printf("%s",string);
    fflush(stdin);
    if( (n = scanf("%lf",&dvalue)) == 1)
    { if((dvalue >= min) && (dvalue <= max))
      {   break;}
    }
    printf("\a");
  }
  return dvalue;        /*  return value */
}
```

Listing 1.4 Get_Double function.

buffer in case it has residual data left from another input operation. Next, we use `scanf` to get one double from the keyboard. If `scanf` returns exactly one value, and if that value is between the minimum and maximum values specified, we break from the loop and return the value.

The `Get_Fixed_String` function is used to get the `descript` string variable in the `main` program. As shown below, this function prompts the user for a string which has a maximum fixed length. The function first prints the prompt string and then a series of underscores equal in number to the maximum length of the string. Next, the cursor is returned to the beginning of the string, and the `Get_String` function is used to get the string from the user. If the string entered is longer than the maximum length, it is simply truncated to the required length. (Certainly, more elegant methods could be used to prompt the user again if the string is too long, but this information is not considered critical and therefore some liberty is taken.)

```
/*=====================================================
  Get_Fixed_String() - prompts user for string of max
  length, underscores fixed length, returns string.
  Prototype:  char *Get_Fixed_String(char *string,
              int len);
  Return:     string of fixed length
  Arguments:  string - a prompt for the user
              len - max length of string
=====================================================*/
char *Get_Fixed_String(char *string,int len)
{ char *fstring;           /*  pointer to fixed string */
  int i;                   /*  loop variable */

  /* prompt user for fixed string,layout underscore */
  printf("%s",string);
  printf("\n\t");
  for(i = 0; i < len ;i++) printf("_");
  for(i = 0; i < len ;i++) printf("\b");
  /* get string, if too long - truncate to max len */
  fstring = Get_String("");
  if((int)strlen(fstring) > len)
  { fstring[len] = '\0';}
  return fstring;      /*  return string */
}
```

Listing 1.5 Get_Fixed_String function.

The final function we'll look at in this chapter is the `Bail_Out` function, which is used to exit the program if an error occurs. The arguments for this function are an error message string and an error number. If the error number is nonzero, the message is printed with a beep and the error number is given before exiting the program. The function can also be used at the end of the program by setting the error number to zero with no message as it is used in this program.

```
/*=========================================================
  Bail_Out() - exits program, if error occurs, sounds
  a beep after printing error message and number
  Prototype:  void Bail_Out(char *message,int number);
  Return:     none
  Arguments:  message - error message to be printed
              number - error number to be printed
=========================================================*/
void Bail_Out(char *message,int number)
{ if(number)
  { printf("\a\n\n\t%s  Error(%i)",message,number);}
  exit(number);
}
```

Listing 1.6 Bail_Out function.

1.5.4 Testing the FILTER Program

Although we haven't discussed filter design in detail, it may be educational to see how the final filter design package will operate. Therefore, let's design a simple filter and view its frequency response. As a first example, we will choose an analog lowpass Butterworth filter with passband and stopband gains of –1 dB and –50 dB, respectively. The passband and stopband edge frequencies should be 500 Hz and 1000 Hz and the data files and frequency plots will be labeled with the title "Lowpass Butterworth Filter." At this point, we can execute the FILTER.EXE file found in the root directory of the software disk. The program can be run directly from the software disk, but for faster operation, it, as well as HELVB.FON, should be copied to the hard disk of a computer. The program is executed by simply typing FILTER at the DOS prompt while in the designated directory.

The first screen that appears is a welcome to the FILTER design program, which will disappear when any key is pressed. Next, we will be asked to enter information about the filter to be designed. Figure 1.7 shows the screen after we have entered the filter parameters for our first example.

(User input is shown in bold italic format.) Once the title for the filter has been entered, the next screen shows the parameters as well as the filter coefficients and gain as shown in Figure 1.8. We haven't learned the significance of these coefficients yet, but they will be discussed in detail in Chapters 2 and 3. The "Print Screen" key can be used to print this screen, or this data can be written to a data file. The name of the file is specified on the next screen.

```
****************** FILTER - The Design Program ******************
***************** Copyright (c) 1994  Les Thede *****************

Please select filter implementation:
       Analog (A)  Dig FIR (F)  Dig IIR (I)
Filter Implementation: A

Please select filter selectivity:
       Lowpass (L)  Highpass (H)  Bandpass (P)  Bandstop (S)
Filter Selectivity: L

Please select filter approximation:
       Butterworth (B)  Chebyshev (C)  Elliptic (E)  Inv-Cheby (I)
Filter Approximation: B

Please enter the passband gain (dB): -1
Please enter the stopband gain (dB): -50
Please enter the passband edge freq (Hz): 500
Please enter the stopband edge freq (Hz): 1000
Please enter a title (40 char max)
       Lowpass Butterworth Filter______________
```

Figure 1.7 Parameter specification screen.

Figure 1.9 shows the next screen which solicits the name for the various data files as well as specifications for the frequency response displays. As indicated on the screen, the user does not have to save the data to files. If no filename is specified, no data files are created. The starting and stopping frequencies are specified for the frequency response which will be plotted on either a linear or logarithmic scale. If the specified frequency range is one decade or greater (a decade is a ratio of 10), the plot will be made using a logarithmic scale, while for frequency ranges of less than a decade, the plot will be made on a linear scale.

```
                         Lowpass Butterworth Filter

                  Filter implementation is:       Analog
                  Filter approximation is:        Butterworth
                  Filter selectivity is:          Lowpass
                  Passband gain          (dB):    -1.00
                  Stopband gain          (dB):    -50.00
                  Passband frequency     (Hz):    500.00
                  Stopband frequency     (Hz):    1000.00

                              Order = 10
                       Overall Gain = 1.00000000e+000

         Numerator Coefficients                  Denominator Coefficients
  [S^2 +        S          +      ()       ]  [S^2 +       S          +      ()      ]
  =================================  =================================
1 0.0 0.0000000e+000 1.12974981e+007  1.0 1.0516073e+003 1.12974981e+007
2 0.0 0.0000000e+000 1.12974981e+007  1.0 3.0518831e+003 1.12974981e+007
3 0.0 0.0000000e+000 1.12974981e+007  1.0 4.7534194e+003 1.12974981e+007
4 0.0 0.0000000e+000 1.12974981e+007  1.0 5.9896579e+003 1.12974981e+007
5 0.0 0.0000000e+000 1.12974981e+007  1.0 6.6395869e+003 1.12974981e+007
```

Figure 1.8 Filter coefficient screen.

The response magnitude can be graphed in decibels or using a linear scale. If the choice of decibels is made, the user can specify the minimum gain to be plotted. Since the exact gain and phase at the filter's edge frequencies may be difficult to determine on the frequency plots, they are calculated and displayed on this screen. As we can see, the passband gain is met, while the stopband gain (–54.34) exceeds the specification.

At this point, pressing any key will bring up the graphics plot of the magnitude frequency response. Another press of a key will display the phase response of the system. These graphs are shown in Figures 1.10 and 1.11. Notice that on the phase plot, the phase angle is always displayed as a value between +180 and –180 degrees. Therefore the two apparent discontinuities are not discontinuities at all.

These plots can be printed to a graphics printer by following a few simple steps. First, the DOS file GRAPHICS.COM must be loaded prior to running FILTER. The type of graphics printer can be specified with GRAPHICS.COM. (See the DOS reference manual for details on how to install GRAPHICS.COM.) After the graphics driver has been installed, FILTER can be executed until the design process reaches the frequency response display stage. Once the graphics screen has been displayed, the screen can be printed by pressing the "Print Screen" key. However, the yellow color of the graph will not show up well on the printout. Therefore, a special feature has been added to allow the yellow graph color to be changed to white which will show up much better on the printout. This feature simply requires the user to press the "P" key (for print) before pressing the "Print Screen" key. This procedure applies to both the magnitude and phase plots.

```
*********************** FILTER - The Design Program **********************
********************** Copyright (c) 1994  Les Thede *********************

Enter a filename (without extension) to save the frequency response and
parameters for this design. Press 'Enter' if no results are to be saved.
Specific extensions as listed below will be used for data files.

   FILE EXTENSION        FILE TYPE AND CONTENTS
   .DTX & .DTB           Text and binary files of params and coefs
   .MAG & .MGB           Text and binary files of magnitude response
   .ANG & .AGB           Text and binary files of phase response

Please enter filename: BUTTER

Please specify response starting freq (Hz): 100
Please specify response stopping freq (Hz): 10000

Please specify response min gain (dB)
(Enter 0 for linear response): -100

********* Edge Frequency Response *********
 Mag(Wp1) =      -1.00  Mag(Ws1) =    -54.34
 Ang(Wp1) =    -404.44  Ang(Ws1) =  -696.77

 Press any key to display magnitude and phase response . . .
```

Figure 1.9 Data file and frequency response screen.

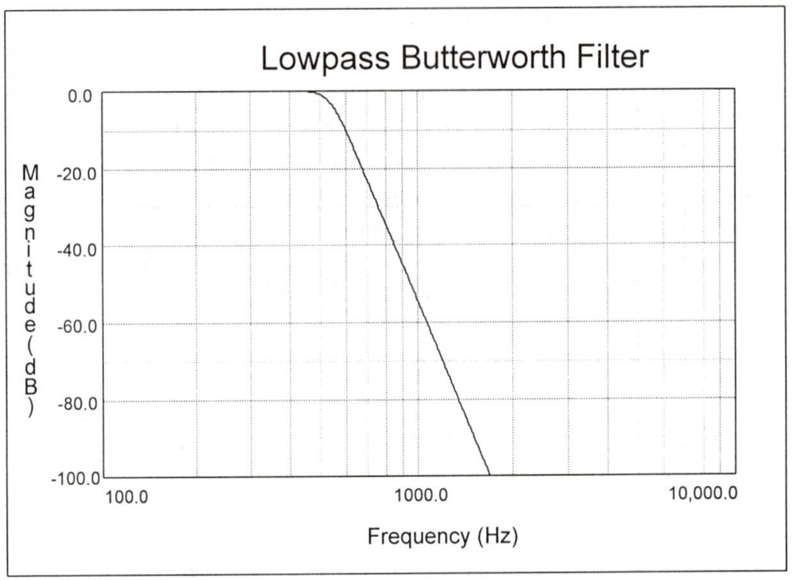

Figure 1.10 Magnitude response screen.

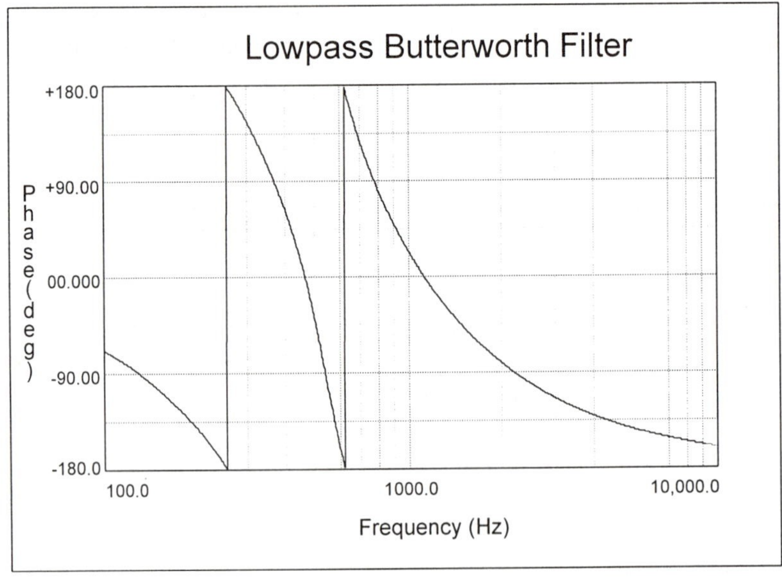

Figure 1.11 Phase response screen.

Following the display of the magnitude and phase plots, the user will be asked whether a new frequency response for the current filter is desired, as indicated in Figure 1.12. This allows the user to change the frequency range for the magnitude and phase plots, or to change the range or mode of the magnitude display. The frequency response data can be displayed in as many ways as desired as long as the user keeps indicating a desire to see a new response. After the final frequency response is viewed, the user will be given a chance to design a new filter. This effectively restarts the FILTER program with a chance to enter all new specifications. As many filters as desired can be designed at a single session.

```
*********************** FILTER - The Design Program **********************
********************** Copyright (c) 1994  Les Thede *********************

Calculate a new freq response for current filter? (Y/N): N

Design a new filter? (Y/N): N
```

Figure 1.12 Continuation screen.

Further details about the FILTER program will be presented in the chapters to come. We will have many more chances to use this program as we develop the theory behind analog and digital filter design.

1.6 Conclusion

At this point we have covered a lot of ground laying the foundation for a successful filter design project. We have studied the standard definitions used to describe filters, as well as the necessary specifications for the proper design of filters. We have developed a framework for a C program which will be used in the design and we have discussed the organization of the software project. We have even taken the final software for a test drive just to become acquainted with the features of the program. But we still need to get under the hood to learn the theory behind filter design and how we can use C to design and implement the filters. We'll begin in the next chapter to study the design of analog filters.

Chapter 2

Analog Filter Approximation Functions

As indicated in the first chapter, an ideal filter is unattainable; the best we can do is to approximate it. There are a number of approximations we can use based on how we want to define "best." In this chapter we will discuss four methods of approximation, each using a slightly different definition. Four sections will be devoted to the major approximation methods used in analog filter design, the Butterworth, Chebyshev, inverse Chebyshev, and elliptic approximations. In each of these sections we will determine the order of the filter required given the filter's specifications and the required normalized transfer function to satisfy the specifications. In the following section we discuss the relative advantages and disadvantages of using these approximation methods. We'll conclude this chapter by writing the necessary code to calculate the filter order and coefficients as well as the code to tie all of these C functions into the framework that we developed in Chapter 1. But first we begin this chapter by describing analog filters mathematically in the form of linear system transfer functions.

2.1 Filter Transfer Functions

An analog filter is a linear system that has an input and output signal. This system's primary purpose is to change the frequency response characteristics of the input signal as it moves through the filter. The characteristics of this filter system could be studied in the time domain or the frequency domain. From a systems point of view, the impulse response $h(t)$ could be used to describe the system in the time domain. The impulse response of a system is the output of a system that has had an impulse applied to the input. Of course, many systems would not be able to sustain an infinite spike (the impulse) being applied to the input of the system, but there are ways to determine $h(t)$ without actually applying the impulse. A filter system can also be described in the frequency domain by using the transfer function $H(s)$. The transfer function of the system can be determined by finding the Laplace transform of $h(t)$. Figure 2.1 indicates that the filter system can be considered either in the time domain or the frequency domain; however, the transfer function description is the predominant method used in filter design, and we will perform most of our filter design using it.

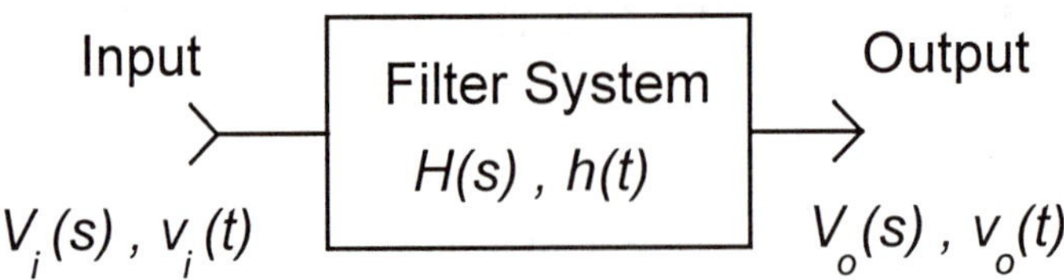

Figure 2.1 The filter as a system.

2.1.1 Transfer Function Characterization

The transfer function *H(s)* for a filter system can be characterized in a number of ways. As shown in Equation 2.1, *H(s)* is typically represented as the ratio of two polynomials in *s* where in this case the numerator polynomial is order m and the denominator is a polynomial of order n. *G* represents an overall gain constant which can take on any value.

$$H(s) = \frac{G \cdot [s^m + a_{m-1} \cdot s^{m-1} + a_{m-2} \cdot s^{m-2} + \cdots + a_1 \cdot s + a_0]}{[s^n + b_{n-1} \cdot s^{n-1} + b_{n-2} \cdot s^{n-2} + \cdots + b_1 \cdot s + b_0]} \tag{2.1}$$

Alternately, the polynomials can be factored to give a form as shown in Equation 2.2. In this representation, the numerator and denominator polynomials have been separated into first-order factors. The *z*'s represent the roots of the numerator and are referred to as the zeros of the transfer function. Similarly, the *p*'s represent the roots of the denominator and are referred to as the poles of the transfer function.

$$H(s) = \frac{G \cdot [(s + z_0) \cdot (s + z_1) \cdots (s + z_{m-1}) \cdot (s + z_m)]}{[(s + p_0) \cdot (s + p_1) \cdots (s + p_{n-1}) \cdot (s + p_n)]} \tag{2.2}$$

Most of the poles and zeros in filter design will be complex valued and will occur as complex conjugate pairs. In this case, it will be more convenient to represent the transfer function as a ratio of quadratic terms which combine the individual complex conjugate factors as shown in Equation 2.3. The first-order factors which are included will be present only if the numerator or denominator polynomial orders are odd. We will be using this form for most of the analog filter design material.

$$H(s) = \frac{G \cdot [(s + z_0) \cdot (s^2 + a_{01} \cdot s + a_{02}) \cdots (s^2 + a_{q1} \cdot s + a_{q2})]}{[(s + p_0) \cdot (s^2 + b_{01} \cdot s + b_{02}) \cdots (s^2 + b_{r1} \cdot s + b_{r2})]} \tag{2.3}$$

As an example of each type of expression, consider the three forms of a transfer function which has a second-order numerator and third-order denominator.

$$H_a(s) = \frac{6.0 \cdot (s^2 + 0.66667)}{s^3 + 3.1650 \cdot s^2 + 5.0081 \cdot s + 4.0001} \tag{2.4a}$$

$$H_b(s) = \frac{6.0 \cdot (s + j0.81650)(s - j0.81650)}{(s + 1.5975) \cdot (s + 0.7837 + j1.3747) \cdot (s + 0.7837 - j1.3747)} \tag{2.4b}$$

$$H_c(s) = \frac{6.0 \cdot (s^2 + 0.66667)}{(s + 1.5975) \cdot (s^2 + 1.5675 \cdot s + 2.5040)} \tag{2.4c}$$

2.1.2 Pole-Zero Plots and Transfer Functions

When the quadratic form of the transfer function is used, it is easy to generate the pole-zero plot for a particular transfer function. The pole-zero plot simply plots the roots of the numerator (zeros) and the denominator (poles) on the complex s-plane. As an example, the pole-zero plot for the sample transfer function given in Equation 2.4 is shown in Figure 2.2.

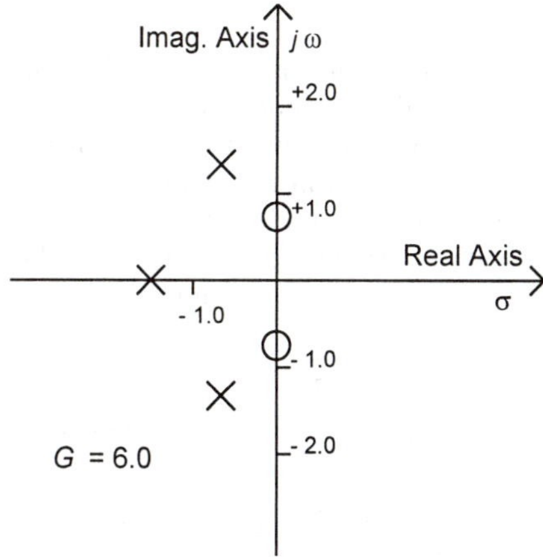

Figure 2.2 Pole-zero plot for Equation 2.4.

Poles are traditionally represented by *X*'s and zeros by *O*'s. If the transfer function is odd, the first-order pole or zero will be located on the real axis. All poles and zeros from the quadratic factors are symmetrically located pairs in the complex plane on opposite sides of the real axis. The gain of the transfer function must be indicated on the plot or the information would be incomplete. Note that there are only two zeros shown, but there is one located at infinity. We can verify this by observing that if we were to allow $|s|$ to approach infinity, $|H(s)|$ would approach zero. Transfer functions always have the same number of poles and zeros, but some exist at infinity.

Conversely, we can also determine a filter's transfer function from the pole-zero plot. In general, any critical frequency (pole or zero) is specified by indicating the real (σ) and imaginary (ω) component. The transfer function would then include a factor of $[s - (\sigma + j\omega)]$. If the critical frequency is complex, we can combine the two complex conjugate factors into a single quadratic factor by multiplying them as shown in Equation 2.5.

$$[s-(\sigma+j\omega)]\cdot[s-(\sigma-j\omega)]=s^2-2\cdot\sigma\cdot s+\sigma^2+\omega^2 \tag{2.5}$$

Example 2.1 Generating a Transfer Function from a Pole-Zero Plot

Problem: Assume that a pole-zero plot shows poles at $(-3 \pm j2)$ and (-4.5) and zeros at $(-5 \pm j1)$ and (-1). Determine the transfer function if its gain is 1.0 at $s = 0$.

Solution: Using the technique of Equation 2.5, the complex conjugate poles and zeros can be combined into quadratic factors as indicated. The first-order factors are handled directly and the gain is included in the numerator. The easiest method to use when given a gain requirement of 1.0 at $s = 0$ is to prepare each factor independently to have a gain of 1 at that frequency as shown in the first transfer function. Then the set of constants can be combined as shown in the second equation.

$$H(s)=\frac{(s^2+10\cdot s+26)}{26}\cdot\frac{(s+1)}{1}\cdot\frac{4.5}{(s+4.5)}\cdot\frac{13}{(s^2+6\cdot s+13)}$$

$$=\frac{2.25\cdot(s+1)\cdot(s^2+10\cdot s+26)}{(s+4.5)\cdot(s^2+6\cdot s+13)}$$

2.1.3 Normalized Transfer Functions

In this chapter we will concentrate on developing what is referred to as a normalized transfer function. A normalized lowpass transfer function is one in which the passband edge radian frequency is set to 1 rad/sec. Of course, this seems a rather unusual frequency, since seldom would a lowpass filter be required to have such a low frequency. However, the technique actually allows the filter designer considerable latitude in designing filters because a normalized transfer function can easily be unnormalized to any other frequency. In the next chapter, we will discuss in detail the procedures used to unnormalize lowpass filters to other frequencies and even learn how to translate a lowpass filter to a highpass, bandpass, or bandstop filter.

Before we begin the development of the approximation functions for analog filters, it may be helpful to go over the general approach taken in these sections. In each case, the general characteristics of the approximation method will be discussed, including its relative advantages and disadvantages. Next, a description of the transfer function for each approximation will be given. There will be no attempt to give an exhaustive derivation of each approximation method in this text, there are more than enough sources of theoretical developments already available. (A list of references for material presented in this chapter is given in Appendix A. The texts by Daniels, Van Valkenburg, and Parks/Burrus are particularly helpful when studying approximation theory.) We will then determine numerical methods to find the order and the coefficients of the transfer function necessary to meet the filter specifications.

2.2 Butterworth Normalized Approximation Functions

The Butterworth approximation function is often called the maximally flat response because no other approximation has a smoother transition through the passband to the stopband. The phase response also is very smooth, which is important when considering distortion. The lowpass Butterworth polynomial has an all-pole transfer function with no finite zeros present. It is the approximation method of choice when low phase distortion and moderate selectivity are required.

2.2.1 Butterworth Approximation Magnitude Response

Equation 2.6 gives the Butterworth approximation's magnitude response where ω_0 is the passband edge frequency for the filter, n is the order of the approximation function, and ε is the passband gain adjustment factor.

The transfer functions will carry subscripts to help identify them in this chapter. In this case, the subscript B indicates a Butterworth filter, and n indicates an nth-order transfer function.

$$\left|H_{B,n}[j(\omega / \omega_o)]\right| = \frac{1}{\sqrt{1+\varepsilon^2 \cdot (\omega / \omega_o)^{2 \cdot n}}} \tag{2.6}$$

where

$$\varepsilon = \sqrt{10^{-0.1 \cdot a_{\text{pass}}} - 1} \tag{2.7}$$

If we set both $\varepsilon = 1$ and $\omega_o = 1$, the filter will have a gain of 1/2 or −3.01 dB at the normalized passband edge frequency of 1 rad/sec.

The Butterworth approximation has a number of interesting properties. First, the response will always have unity gain at $\omega = 0$, no matter what value is given to ε. However, the gain at the normalized passband edge frequency of $\omega = 1$ will depend on the value of ε. In addition, the response gain decreases by a factor of $-20n$ dB per decade of frequency change. That happens because for large ω, the transfer function gain becomes inversely proportional to ω which increases by 10 for every decade. (A decade in frequency is a ratio of 10. For example, the span of frequencies from 1 to 10 rad/sec and the span of frequencies from 1000 to 10,000 Hz are both referred to as one decade.) Therefore, if we design a fifth-order Butterworth filter, the gain will decrease 100 dB per decade for frequencies above the passband edge frequency.

2.2.2 Butterworth Approximation Order

The order of the Butterworth filter is dependent on the specifications provided by the user. These specifications include the edge frequencies and gains. The standard formula for the Butterworth order calculation is given in Equation 2.8. In this formulation, note that it is the ratio of the stopband and passband frequencies which is important, not either one of these independently. This means that a filter with a given set of gains will require the same order whether the edge frequencies are 100 and 200 rad/sec or 100,000 and 200,000 Hz. The value of n calculated using this equation must always be rounded to the next highest integer in order to guarantee that the specifications will be met by the integer order of the filter designed.

$$n_B = \frac{\log[(10^{-0.1 \cdot a_{\text{stop}}} - 1) / (10^{-0.1 \cdot a_{\text{pass}}} - 1)]}{2 \cdot \log(\omega_{\text{stop}} / \omega_{\text{pass}})} \tag{2.8}$$

2.2.3 Butterworth Approximation Pole Locations

The poles for a Butterworth approximation function are equally spaced around a circle in the s-plane and are symmetrical about the $j\omega$ axis. Plotting the poles of the magnitude-squared function $|H(s)|^2$ shows twice as many poles as the order of the filter. We are able to determine the Butterworth transfer function from the poles in the left half plane (LHP) which produce a stable system. In order to determine the exact pole positions in the s-plane we use the polar form for specifying the complex location. For each of the poles, we must know the distance from the origin (the radius of the circle) and the angle from the positive real axis.

The radius of the circle for our normalized case is a function of the passband gain and is given in Equation 2.9.

$$R = \varepsilon^{-1/n} \tag{2.9}$$

Once the radius of the circle is known, the pole positions are determined by calculating the necessary angles. Equation 2.10 can be used to determine the angles for those complex poles in the second quadrant.

$$\theta_m = \frac{\pi \cdot (2 \cdot m + n + 1)}{2 \cdot n}, \quad m = 0,\ 1,\ \ldots,\ (n/2) - 1 \quad (n \text{ even}) \tag{2.10a}$$

$$\theta_m = \frac{\pi \cdot (2 \cdot m + n + 1)}{2 \cdot n}, \quad m = 0,\ 1,\ \ldots,\ [(n-1)/2] - 1 \quad (n \text{ odd}) \tag{2.10b}$$

It is important to remember that in this equation θ_m represents only the angles in the second quadrant *which have complex conjugates in the third quadrant*. In other words, θ_m does *not* include the pole on the real axis for odd-order functions. For this reason, Equation 2.10b is valid only for odd-order filters where $n \geq 3$ since a first-order filter would have no complex conjugate poles. (We'll see that this definition allows a cleaner algorithm for the C code which we will write later.) The precise pole locations can then be determined from Equations 2.11 and 2.12.

$$\sigma_m = R \cdot \cos(\theta_m) \quad (2.11)$$

$$\omega_m = R \cdot \sin(\theta_m) \quad (2.12)$$

In the case of odd-order transfer functions, the first-order pole will be located at a position σ_R equal to the radius of the circle as indicated in Equation 2.13.

$$\sigma_R = -R \quad (2.13)$$

2.2.4 Butterworth Approximation Transfer Functions

The Butterworth transfer function can be determined from the pole locations in the LHP as we have seen in the first section of this chapter. Since most of these poles are complex conjugate pairs (except for the possible pole on the real axis for odd orders), we can get all of the information we need from the poles in the second quadrant. The complete approximation transfer function can be determined from a combination of a first-order factor (for odd orders) and quadratic factors. Each of these factors will have a constant in the numerator to adjust the gain to unity at $\omega = 0$ as illustrated in Example 2.1. We will start by defining the form of the first-order factor in Equation 2.14. Note that at this point the transfer function variables are uppercase *S*'s where prior to this we have been using lowercase *s*'s. This is an attempt to distinguish between the normalized transfer function (using *S*'s) and the unnormalized functions (indicated by *s*'s) which will be developed in the next chapter.

$$H_o(S) = \frac{R}{S + R} \quad (2.14)$$

For each complex conjugate pole in the second quadrant, there will be the following quadratic factor in the transfer function.

$$H_m(S) = \frac{B_{2m}}{S^2 + B_{1m} \cdot S + B_{2m}} \quad (2.15)$$

where

$$B_{1m} = -2 \cdot \sigma_m \tag{2.16}$$

$$B_{2m} = \sigma_m^2 + \omega_m^2 \tag{2.17}$$

The complete Butterworth transfer function can now be defined as shown in Equations 2.18.

$$H_{B,n}(S) = \frac{\prod_m (B_{2m})}{\prod_m (S^2 + B_{1m} \cdot S + B_{2m})}, \tag{2.18a}$$
$$m = 0, 1, \text{K} , (n/2) - 1 \ (n \text{ even})$$

$$H_{B,n}(S) = \frac{R \cdot \prod_m (B_{2m})}{(S+R) \cdot \prod_m (S^2 + B_{1m} \cdot S + B_{2m})}, \tag{2.18b}$$
$$m = 0, 1, \text{K} , [(n-1)/2] - 1 \ (n \text{ odd})$$

We have now reached a point where some examples are in order. First, we will consider some numerical examples, and then we will test the FILTER program on the same specifications.

Example 2.2 Butterworth Third-Order Normalized Transfer Function

Problem: Determine the order, pole locations, and transfer function coefficients for a Butterworth filter to satisfy the following specifications:

$a_{pass} = -1$ dB, $a_{stop} = -12$ dB, $\omega_{pass} = 1$ rad/sec, and $\omega_{stop} = 2$ rad/sec

Solution: First, we determine the fundamental constants needed from Equations 2.7 – 2.9.

$\varepsilon = 0.508847$ $\qquad n = 2.92$ (3rd order) $\qquad R = 1.252576$

Next, we find the locations of the first-order pole and the complex pole in the second quadrant from Equations 2.10 – 2.13.

(1st order)	$\sigma_R = -1.252576$	$\omega_R = 0.0$
$\theta_0 = 2\pi/3$	$\sigma_0 = -0.626288$	$\omega_0 = +1.084763$

And, finally, we generate the transfer function from Equations 2.14 – 2.18.

$$H_{B,3}(S) = \frac{1.252576 \cdot 1.568948}{(S + 1.252576) \cdot (S^2 + 1.252576 \cdot S + 1.568948)}$$

Example 2.3 Butterworth Fourth-Order Normalized Transfer Function

Problem: Determine the order, pole locations, and transfer function coefficients for a Butterworth filter to satisfy the following specifications:

$a_{\text{pass}} = -1$ dB, $a_{\text{stop}} = -18$ dB, $\omega_{\text{pass}} = 1$ rad/sec, and $\omega_{\text{stop}} = 2$ rad/sec

Solution: First, we determine the fundamental constants needed from Equations 2.7 – 2.9.

$\varepsilon = 0.508847$	$n = 3.95$ (4th order)	$R = 1.184004$

Next, we find the locations of the two complex poles in the second quadrant from Equations 2.10 – 2.13.

$\theta_0 = 5\pi/8$	$\sigma_0 = -0.453099$	$\omega_0 = +1.093877$
$\theta_1 = 7\pi/8$	$\sigma_1 = -1.093877$	$\omega_1 = +0.453099$

And, finally, we generate the transfer function from Equations 2.14 – 2.18.

$$H_{B,4}(S) = \frac{(1.401866)^2}{(S^2 + 2.18775 \cdot S + 1.40187) \cdot (S^2 + 0.90620 \cdot S + 1.40187)}$$

In order to use FILTER to determine the normalized transfer functions, we will assume a passband edge frequency of 1 rad/sec (0.159154943092 Hz) and a stopband edge frequency of 2 rad/sec (0.318309886184 Hz). (We must enter the frequencies into FILTER using the Hertz values and they must have more significant digits than we require in our answer.) Twelve significant digits were used to enter the edge frequencies, but they are displayed on the coefficient screen with only two significant digits. Rest assured that they are stored internally with the higher accuracy, but the display is set for the more typical requirements of filter frequency. The coefficient values determined are shown in Figures 2.3 and 2.4.

```
                    Butterworth 3rd Order Lowpass

               Filter implementation is:       Analog
               Filter approximation is:        Butterworth
               Filter selectivity is:          Lowpass
               Passband gain          (dB):    -1.00
               Stopband gain          (dB):    -12.00
               Passband frequency     (Hz):    0.16
               Stopband frequency     (Hz):    0.32

                          Order = 3
                    Overall Gain = 1.00000000e+000

      Numerator Coefficients                 Denominator Coefficients
  [S^2 +       S          +     ()      ]  [S^2 +       S         +     ()      ]
  ==================================  ==================================
1 0.0 0.00000000e+00 1.25257639e+000  0.0 1.00000000e+00 1.25257639e+000
2 0.0 0.00000000e+00 1.56894761e+000  1.0 1.25257639e+00 1.56894761e+000
```

Figure 2.3 Butterworth normalized third-order coefficients.

```
                    Butterworth 4th Order Lowpass

               Filter implementation is:       Analog
               Filter approximation is:        Butterworth
               Filter selectivity is:          Lowpass
               Passband gain          (dB):    -1.00
               Stopband gain          (dB):    -18.00
               Passband frequency     (Hz):    0.16
               Stopband frequency     (Hz):    0.32

                          Order = 4
                    Overall Gain = 1.00000000e+000

      Numerator Coefficients                 Denominator Coefficients
  [S^2 +       S          +     ()      ]  [S^2 +       S         +     ()      ]
  ==================================  ==================================
1 0.0 0.00000000e+00 1.40186545e+000  1.0 9.06197421e-01 1.40186545e+000
2 0.0 0.00000000e+00 1.40186545e+000  1.0 2.18775410e+00 1.40186545e+000
```

Figure 2.4 Butterworth normalized fourth-order coefficients.

The associated magnitude and phase responses are shown in Figures 2.5 and 2.6 and illustrate the difference between a third-order and fourth-order filter. Notice that the magnitude response uses a different scale for the passband and stopband response. (FILTER cannot put two different responses on the same graph or use different scales for passband and stopband — sorry. They are displayed here in that manner for ease of comparison. However, the magnitude scale can be changed on different graphs to provide better detail.)

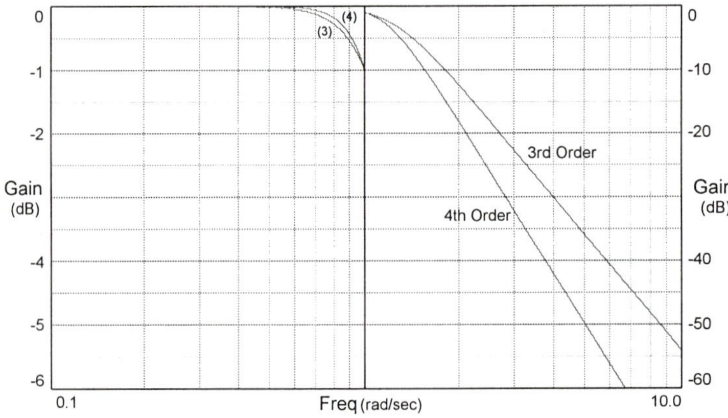

Figure 2.5 Butterworth third-order and fourth-order magnitude responses.

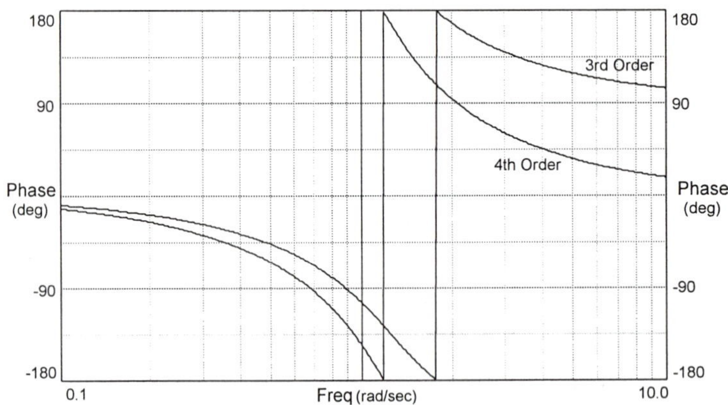

Figure 2.6 Butterworth third-order and fourth-order phase responses.

2.3 CHEBYSHEV NORMALIZED APPROXIMATION FUNCTIONS

The Chebyshev approximation function also has an all-pole transfer function like the Butterworth approximation. However, unlike the Butterworth case, the Chebyshev filter allows variation or ripple in the passband of the filter. This reduction in the restrictions placed on the characteristics of the passband enables the transition characteristics of the Chebyshev to be steeper than the Butterworth transition. Because of this more rapid transition, the Chebyshev filter is able to satisfy user specifications with lower-order filters than the Butterworth case. However, the phase response is not as linear as the Butterworth case, and therefore if low phase distortion is a priority, the Chebyshev approximation may not be the best choice.

2.3.1 Chebyshev Approximation Magnitude Response

The magnitude response function for the Chebyshev approximation is shown in Equation 2.19.

$$\left|H_{C,n}[j(\omega/\omega_o)]\right| = \frac{1}{\sqrt{1+\varepsilon^2 \cdot C_n^2(\omega/\omega_o)}} \tag{2.19}$$

where the definition of ε is again

$$\varepsilon = \sqrt{10^{-0.1 \cdot a_{\text{pass}}} - 1} \tag{2.20}$$

and $C_n(\omega)$ is the Chebyshev polynomial of the first kind of degree n. The normalized Chebyshev polynomial ($\omega_o = 1$) is defined as

$$C_n(\omega) = \cos[n \cdot \cos^{-1}(\omega)], \quad \omega \leq 0 \tag{2.21a}$$

$$C_n(\omega) = \cosh[n \cdot \cosh^{-1}(\omega)], \quad \omega > 0 \tag{2.21b}$$

We can see that the mathematical description used for this approximation is more involved than the Butterworth case. We will be concerned with the expression where $\omega > 0$, but the Chebyshev polynomial has many interesting features which are discussed in the references listed at the end of this text.

2.3.2 Chebyshev Approximation Order

The order of the Chebyshev filter will be dependent on the specifications provided by the user. The general form of the calculation for the order is the same as for the Butterworth, except that the inverse hyperbolic cosine function is used in place of the common logarithm function. As in the Butterworth case, the value of n actually calculated must be rounded to the next highest integer in order to guarantee that the specifications will be met.

$$n_C = \frac{\cosh^{-1}\left[\sqrt{(10^{-0.1 \cdot a_{\text{stop}}} - 1) / (10^{-0.1 \cdot a_{\text{pass}}} - 1)}\right]}{\cosh^{-1}(\omega_{\text{stop}} / \omega_{\text{pass}})} \tag{2.22}$$

2.3.3 Chebyshev Approximation Pole Locations

The poles for a Chebyshev approximation function are located on an ellipse instead of a circle as in the Butterworth case. The ellipse is centered at the origin of the s-plane with its major axis oriented along the $j\omega$ axis with intercepts of $\pm \cosh(D)$, while the minor axis is oriented along the real axis with intercepts of $\pm \sinh(D)$. The variable D is defined as

$$D = \frac{\sinh^{-1}(\varepsilon^{-1})}{n} \tag{2.23}$$

The pole locations can be defined in terms of D and an angle ϕ as shown in Equation 2.24. The angles determined locate the poles of the transfer function in the first quadrant. However, we can use them to find the poles in the second quadrant by simply changing the sign of the real part of each complex pole. The real and imaginary components of the pole locations can now be defined as shown in Equations 2.25 and 2.26.

$$\phi_m = \frac{\pi \cdot (2 \cdot m + 1)}{2 \cdot n}, \quad m = 0,\ 1,\ \text{K}\ ,\ (n/2) - 1 \ \ (n \text{ even}) \tag{2.24a}$$

$$\phi_m = \frac{\pi \cdot (2 \cdot m + 1)}{2 \cdot n}, \quad m = 0,\ 1,\ \text{K}\ ,\ [(n-1)/2] - 1 \ \ (n \text{ odd}) \tag{2.24b}$$

$$\sigma_m = -\sinh(D) \cdot \sin(\phi_m) \tag{2.25}$$

$$\omega_m = \cosh(D) \cdot \cos(\phi_m) \tag{2.26}$$

If the function has an odd-order, there will be a real pole located in the LHP as indicted by Equation 2.27.

$$\sigma_R = -\sinh(D) \tag{2.27}$$

2.3.4 Chebyshev Approximation Transfer Functions

Using the results of Equation 2.27, we know that an odd-order Chebyshev transfer function will have a factor of the form illustrated in Equation 2.28.

$$H_o(S) = \frac{\sinh(D)}{S + \sinh(D)} \tag{2.28}$$

The quadratic factors for the Chebyshev transfer function will take on exactly the same form as the Butterworth case as shown below.

$$H_m(S) = \frac{B_{2m}}{S^2 + B_{1m} \cdot S + B_{2m}} \tag{2.29}$$

$$B_{1m} = -2 \cdot \sigma_m \tag{2.30}$$

$$B_{2m} = \sigma_m^2 + \omega_m^2 \tag{2.31}$$

We are now just about ready to define the general form of the Chebyshev transfer function. However, one small detail still must be considered. Because there is ripple in the passband, Chebyshev even and odd-order approximations do *not* have the same gain at $\omega = 0$. As seen in Figure 2.9, each approximation has a number of half-cycles of ripple in the passband equal to the order of the filter. This forces even-order filters to have a gain of a_{pass} at $\omega = 0$. However, the first-order and quadratic factors we have defined are all set to give 0 dB gain at $\omega = 0$. Therefore, if no adjustment of gain is made to even-order Chebyshev approximations, they would have a gain of 0 dB at $\omega = 0$ and a gain of $-a_{pass}$ (that is, a gain greater than 1.0) at certain other frequencies where the ripple peaks. A gain constant must therefore be included for even-order transfer functions with the value of

$$G = 10^{0.05 \cdot a_{pass}} \tag{2.32}$$

We are now ready to define a generalized transfer function for the Chebyshev approximation function as shown below.

$$H_{C,n}(S) = \frac{(10^{0.05 \cdot a_{pass}}) \cdot \prod_m (B_{2m})}{\prod_m (S^2 + B_{1m} \cdot S + B_{2m})}, \tag{2.33a}$$

$$m = 0,\ 1,\ \ldots,\ (n/2) - 1 \ \ (n \text{ even})$$

$$H_{C,n}(S) = \frac{\sinh(D) \cdot \prod_m (B_{2m})}{(S + \sinh(D)) \cdot \prod_m (S^2 + B_{1m}S + B_{2m})}, \tag{2.33b}$$

$$m = 0,\ 1,\ \ldots,\ [(n-1)/2] - 1 \ \ (n \text{ odd})$$

It is again time to consider some numerical examples before using FILTER to determine the filter coefficients.

Example 2.4 Chebyshev Third-Order Normalized Transfer Function

Problem: Determine the order, pole locations, and coefficients of the transfer function for a Chebyshev filter to satisfy the following specifications:

$a_{pass} = -1$ dB, $a_{stop} = -22$ dB, $\omega_{pass} = 1$ rad/sec, and $\omega_{stop} = 2$ rad/sec

Solution: First, we determine the fundamental constants needed from Equations 2.20, 2.22, and 2.23.

$\varepsilon = 0.508847$ $\quad n = 2.96$ (3rd order)

$D = 0.475992$ $\quad \cosh(D) = 1.115439$ $\quad \sinh(D) = 0.494171$

Next, we find the locations of the first-order pole and the complex pole in the second quadrant from Equations 2.24 – 2.27.

(1st order)	$\sigma_R = -0.494171$	$\omega_R = 0.0$
$\phi_0 = 1\pi/6$	$\sigma_0 = -0.247085$	$\omega_0 = +0.965999$

And, finally, we generate the transfer function from Equations 2.28 – 2.33.

$$H_{C,3}(s) = \frac{0.494171 \cdot 0.994205}{(S + 0.494171) \cdot (S^2 + 0.494171 \cdot S + 0.994205)}$$

Example 2.5 Chebyshev Fourth-Order Normalized Transfer Function

Problem: Determine the order, pole locations, and transfer function coefficients for a Chebyshev filter to satisfy the following specifications:

$a_{pass} = -1$ dB, $a_{stop} = -33$ dB, $\omega_{pass} = 1$ rad/sec, and $\omega_{stop} = 2$ rad/sec

Solution: First, we determine the fundamental constants needed from Equations 2.20, 2.22, and 2.23.

$\varepsilon = 0.508847$	$n = 3.92$ (4th order)	
$D = 0.356994$	$\cosh(D) = 1.064402$	$\sinh(D) = 0.364625$

Next, we find the locations of the two complex poles in the second quadrant from Equations 2.24 – 2.27.

$\theta_0 = 1\pi/8$	$\sigma_0 = -0.139536$	$\omega_0 = +0.983379$
$\theta_1 = 3\pi/8$	$\sigma_1 = -0.336870$	$\omega_1 = +0.407329$

And, finally, we generate the transfer function from Equations 2.28 – 2.33. Note that in this even-order case, the gain constant of 0.891251 is included.

$$H_{C,4}(S) = \frac{0.891251 \cdot 0.986505 \cdot 0.279398}{(S^2 + 0.27907 \cdot S + 0.98651) \cdot (S^2 + 0.67374 \cdot S + 0.27940)}$$

The results of running FILTER for these Chebyshev specifications are shown in Figures 2.7 and 2.8. The magnitude and phase responses for the third and fourth-order Chebyshev filters are shown in Figures 2.9 and 2.10.

```
                  Chebyshev 3rd Order Lowpass

            Filter implementation is:     Analog
            Filter approximation is:      Chebyshev
            Filter selectivity is:        Lowpass
            Passband gain         (dB):   -1.00
            Stopband gain         (dB):   -22.00
            Passband frequency    (Hz):   0.16
            Stopband frequency    (Hz):   0.32

                         Order = 3
                  Overall Gain = 1.00000000e+000

     Numerator Coefficients              Denominator Coefficients
  [S^2 +      S          +      ()     ]  [S^2 +      S          +      ()     ]
  ==================================  ==================================
1 0.0 0.00000000e+00 4.94170605e-001  0.0 1.0000000e+000 4.94170605e-001
2 0.0 0.00000000e+00 9.94204587e-001  1.0 4.94170605e-01 9.94204587e-001
```

Figure 2.7 Chebyshev normalized third-order coefficients.

```
                  Chebyshev 4th Order Lowpass

            Filter implementation is:     Analog
            Filter approximation is:      Chebyshev
            Filter selectivity is:        Lowpass
            Passband gain         (dB):   -1.00
            Stopband gain         (dB):   -33.00
            Passband frequency    (Hz):   0.16
            Stopband frequency    (Hz):   0.32

                         Order = 4
                  Overall Gain = 8.91250938e-001

     Numerator Coefficients              Denominator Coefficients
  [S^2 +      S          +      ()     ]  [S^2 +      S          +      ()     ]
  ==================================  ==================================
1 0.0 0.00000000e+00 9.86504875e-001  1.0 2.79071992e-01 9.86504875e-001
2 0.0 0.00000000e+00 2.79398094e-001  1.0 6.73739388e-01 2.79398094e-001
```

Figure 2.8 Chebyshev normalized fourth-order coefficients.

2.4 Inverse Chebyshev Normalized Approximation Functions

The inverse Chebyshev approximation function, also called the Chebyshev type II function, is a rational approximation with both poles and zeros in its transfer function. This approximation has a smooth, maximally flat response in the passband, just as the Butterworth approximation, but has ripple in the stopband caused by the zeros of the transfer function. The inverse Chebyshev provides better transition characteristics than the Butterworth and better phase response than the standard Chebyshev. Although the inverse Chebyshev has these beneficial features, it is more involved to design.

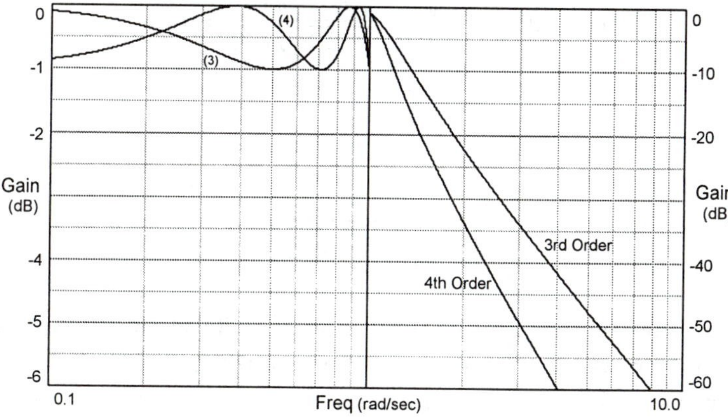

Figure 2.9 Chebyshev third-order and fourth-order magnitude responses.

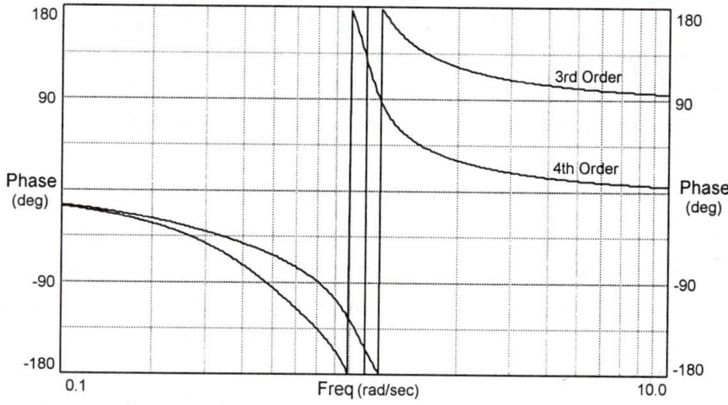

Figure 2.10 Chebyshev third-order and fourth-order phase responses.

2.4.1 Inverse Chebyshev Approximation Magnitude Response

The development of the inverse Chebyshev response is derived from the standard Chebyshev response. We will discuss the methods needed to determine the inverse Chebyshev approximation function while leaving the intricate details to the reference works. The name "inverse Chebyshev" is well-deserved in this case since we will see that many of the computations are based on inverse or opposite values from the standard computations. Let's begin with the definition of the magnitude frequency response function as shown in Equation 2.34

$$\left|H_{I,n}[j(\omega / \omega_o)]\right| = \frac{\sqrt{\varepsilon_i^2 \cdot C_n^2(\omega_o / \omega)}}{\sqrt{1+\varepsilon_i^2 \cdot C_n^2(\omega_o / \omega)}} \tag{2.34}$$

where

$$\varepsilon_i = \frac{1}{\sqrt{10^{-0.1 \cdot a_{\text{stop}}} - 1}} \tag{2.35}$$

The first observation concerning Equation 2.34 is that it indeed has a numerator portion which allows for the finite zeros in the transfer function. Upon closer inspection, we find the use of ε_i in place of ε. Equation 2.35 indicates ε_i is the inverse of ε and a_{pass} is replaced with a_{stop}. Because of the differences, we will use the subscript to distinguish ε_i from the standard ε. Although C_n still represents the Chebyshev polynomial of the first kind of degree n as defined in Equation 2.21, we notice that the argument of the function is the inverse of the standard definition (ω_o/ω instead of ω/ω_o). We will see a little later in this section how these differences affect our determination of the poles and zeros of the transfer function.

2.4.2 Inverse Chebyshev Approximation Order

Because of the nature of the derivation of the inverse Chebyshev approximation function from the standard Chebyshev approximation, it should come as no surprise that the calculation of the order for an inverse Chebyshev is the same as for the standard Chebyshev. The expression is given in Equation 2.36 and is the same as Equation 2.22 except for the subscript I designating the calculation as the inverse Chebyshev order.

$$n_I = \frac{\cosh^{-1}\left[\sqrt{(10^{-0.1 \cdot a_{\text{stop}}} - 1) / (10^{-0.1 \cdot a_{\text{pass}}} - 1)}\right]}{\cosh^{-1}(\omega_{\text{stop}} / \omega_{\text{pass}})} \tag{2.36}$$

2.4.3 Inverse Chebyshev Approximation Pole-Zero Locations

The determination of the pole locations for the normalized inverse Chebyshev approximation is based on techniques similar to those used for the standard normalized Chebyshev approximation. The pole positions for the inverse Chebyshev case are found using the same values of ϕ_m, but the value of ε_i is calculated differently. Once the pole positions are found however, the inverse Chebyshev poles are the reciprocals of the standard poles. (There's that inverse relationship again.) For example, if there exists a standard Chebyshev pole at

$$p = \sigma + j\omega \tag{2.37}$$

then the reciprocal of p gives the inverse Chebyshev pole position as

$$p^{-1} = \frac{\sigma - j\omega}{(\sigma + j\omega)\cdot(\sigma - j\omega)} = \frac{\sigma}{\sigma^2 + \omega^2} - j\frac{\omega}{\sigma^2 + \omega^2} \tag{2.38}$$

Notice that if a pole's distance from the origin is greater than one, the reciprocal's distance will be less than one, and vice versa. In addition, the position of the pole is reflected across the real axis, so although the original pole position may be in the second quadrant, the reciprocal is located in the third quadrant. Consequently, if we are able to determine pole positions for the standard Chebyshev approximation function as discussed in the previous section, we should have little problem finding the inverse Chebyshev pole locations.

Let's derive the mathematical equations necessary to determine the pole locations for the inverse Chebyshev approximation function along the same lines as we did for the standard Chebyshev case. First, D_i will be defined in terms of ε_i in Equation 2.39.

$$D_i = \frac{\sinh^{-1}(\varepsilon_i^{-1})}{n} \tag{2.39}$$

Next, we can define the pole locations in the second quadrant in the manner of the previous section as shown in Equations 2.40 – 2.42, remembering that these primed values must still be inverted.

$$\sigma'_m = -\sinh(D_i)\cdot\sin(\phi_m) \tag{2.40}$$

$$\omega'_m = \cosh(D_i)\cdot\cos(\phi_m) \tag{2.41}$$

$$\phi_m = \frac{\pi\cdot(2\cdot m+1)}{2\cdot n},\ m=0,\ 1,\ \ldots,\ (n/2)-1\ \ (n\text{ even}) \tag{2.42a}$$

$$\phi_m = \frac{\pi\cdot(2\cdot m+1)}{2\cdot n},\ m=0,\ 1,\ \ldots,\ [(n-1)/2]-1\ \ (n\text{ odd}) \tag{2.42b}$$

We can determine the final pole locations by inverting these poles as indicated in Equations 2.43 and 2.44.

$$\sigma_m = \frac{\sigma'_m}{\sigma'^2_m + \omega'^2_m} \tag{2.43}$$

$$\omega_m = \frac{-\omega'_m}{\sigma'^2_m + \omega'^2_m} \tag{2.44}$$

If the approximation function is odd-order, then there will be a first-order pole on the negative real axis at σ_R as defined in Equation 2.45.

$$\sigma_R = -[\sinh(D_i)]^{-1} \tag{2.45}$$

Next we need to determine the placement of the finite zeros of the inverse Chebyshev approximation function which are all purely imaginary complex conjugate pairs located on the $j\omega$ axis. Because they only occur in pairs, the numerator of an inverse Chebyshev transfer function will always be even. If the order of the denominator is odd, then one zero of the transfer function will be located at infinity. The location of the zeros on the $j\omega$ axis is determined by Equations 2.46 and 2.47 where ϕ_m is as defined in Equation 2.42. A z is used in the subscript to differentiate the zero locations from the pole locations. (By the way, did you notice that the secant function in 2.47 is the *inverse* of the cosine function used in the standard Chebyshev function?)

$$\sigma_{zm} = 0.0 \tag{2.46}$$

$$\omega_{zm} = \sec(\phi_m) \tag{2.47}$$

2.4.4 Inverse Chebyshev Approximation Transfer Functions

Now that we have located the necessary poles and zeros which are pertinent to the definition of the inverse Chebyshev approximation, we can define the various factors which describe the transfer function. First, for odd-order approximations, Equation 2.48 describes the first-order factor.

$$H_o(S) = \frac{[\sinh(D_i)]^{-1}}{S + [\sinh(D_i)]^{-1}} \tag{2.48}$$

Next, the quadratic components of the transfer function are described in Equations 2.49 – 2.53. These are similar to the quadratic definition for the Chebyshev case, but we have added a numerator quadratic for the zeros as well.

$$H_m(S) = \frac{B_{2m} \cdot (S^2 + A_{1m} \cdot S + A_{2m})}{A_{2m} \cdot (S^2 + B_{1m} \cdot S + B_{2m})} \tag{2.49}$$

where

$$B_{1m} = -2 \cdot \sigma_m \tag{2.50}$$

$$B_{2m} = \sigma_m^2 + \omega_m^2 \tag{2.51}$$

$$A_{1m} = -2 \cdot \sigma_{zm} = 0.0 \tag{2.52}$$

$$A_{2m} = \sigma_{zm}^2 + \omega_{zm}^2 = \omega_{zm}^2 \tag{2.53}$$

Although the value of A_{1m} is zero, it is included to be consistent with the format which will be used throughout the remainder of the text.

We are now ready to define the generalized transfer function form for the inverse Chebyshev approximation function which is shown in Equation 2.54. Since the inverse Chebyshev has a maximally flat response in the passband as the Butterworth, there is no need for a gain adjustment constant as in the standard Chebyshev case.

$$H_{I,n}(S) = \frac{\prod_m (B_{2m}) \cdot \prod_m (S^2 + A_{1m} \cdot S + A_{2m})}{\prod_m (A_{2m}) \cdot \prod_m (S^2 + B_{1m} \cdot S + B_{2m})}, \qquad (2.54a)$$

$$m = 0,\ 1,\ \mathrm{K}\ ,\ (n/2) - 1 \ \ (n \text{ even})$$

$$H_{I,n}(S) = \frac{[\sinh(D_i)]^{-1} \cdot \prod_m (B_{2m}) \cdot \prod_m (S^2 + A_{1m} \cdot S + A_{2m})}{(S + [\sinh(D_i)]^{-1}) \cdot \prod_m (A_{2m}) \cdot \prod_m (S^2 + B_{1m} \cdot S + B_{2m})}, \qquad (2.54b)$$

$$m = 0,\ 1,\ \mathrm{K}\ ,\ [(n-1)/2] - 1 \ \ (n \text{ odd})$$

The following numerical examples should help to illustrate the process.

Example 2.6 Inv. Chebyshev Third-Order Normal. Transfer Function

Problem: Determine the order, pole and zero locations, and transfer function coefficients for an inverse Chebyshev filter to satisfy the following specifications:

$a_{\text{pass}} = -1$ dB, $a_{\text{stop}} = -22$ dB, $\omega_{\text{pass}} = 1$ rad/sec, and $\omega_{\text{stop}} = 2$ rad/sec

Solution: First, we determine the fundamental constants needed from Equations 2.35, 2.36, and 2.39.

$\varepsilon_i = 0.079685$	$n = 2.96$ (3rd order)	
$D_i = 1.074803$	$\cosh(D_i) = 1.635391$	$\sinh(D_i) = 1.294026$

Next, we find the locations of the first-order pole, the complex pole in the second quadrant, and the second-order zeros on the $j\omega$ axis from Equations 2.40 – 2.47.

(1st order)	$\sigma_R = -0.772782$	$\omega_R = 0.0$
$\phi_0 = 1\pi/6$	$\sigma'_0 = -0.647013$	$\omega'_0 = +1.416290$
	$\sigma_0 = -0.266864$	$\omega_0 = -0.584157$
(zeros)	$\sigma_{z0} = +0.0$	$\omega_{z0} = +1.154701$

And finally, we generate the transfer function from Equations 2.48 – 2.54.

$$H^*_{I,3}(S) = \frac{0.772782 \cdot 0.412456 \cdot (S^2 + 1.333333)}{1.333333 \cdot (S + 0.772782) \cdot (S^2 + 0.533728 \cdot S + 0.412456)}$$

$$H_{I,3}(S) = \frac{0.478108 \cdot (S^2 + 5.333333)}{(S + 1.545564) \cdot (S^2 + 1.067457 \cdot S + 1.649823)}$$

There is a problem with the first transfer function (shown with an asterisk *) which we have just developed. It implements an inverse Chebyshev approximation function which is normalized to $\omega_{stop} = 1$ rad/sec instead of $\omega_{pass} = 1$ rad/sec. This means that the entire frequency response is a factor of 2 too low. The attenuation at $\omega = 0.5$ rad/sec is ~ 1 dB and the attenuation at $\omega = 1$ rad/sec ~22 dB. The process we can use to correct the problem is actually an unnormalization procedure which is covered in Chapter 3. This unnormalization will usually occur as part of the total filter design process, but we can make the adjustment manually in this particular case. The correct transfer function can be determined by substituting *S*/2 for *S* and then simplifying as indicated in the second transfer function.

Example 2.7 Inv. Chebyshev Fourth-Order Normal. Transfer Function

Problem: Determine the order, pole and zero locations, and transfer function coefficients for an inverse Chebyshev filter to satisfy the following specifications:

$a_{pass} = -1$ dB, $a_{stop} = -33$ dB, $\omega_{pass} = 1$ rad/sec, and $\omega_{stop} = 2$ rad/sec

Solution: First, we determine the fundamental constants needed from Equations 2.35, 2.36, and 2.39.

$\varepsilon_i = 0.022393$	$n = 3.92$ (4th order)	
$D_i = 1.123072$	$\cosh(D_i) = 1.699781$	$\sinh(D_i) = 1.374502$

Next, we find the locations of the two complex poles in the second quadrant and the second-order zeros from Equations 2.40 – 2.47.

$\phi_0 = 1\pi/8$	$\sigma'_0 = -0.525999$	$\omega'_0 = +1.570393$
	$\sigma_0 = -0.191774$	$\omega_0 = -0.572549$
$\phi_0 = 3\pi/8$	$\sigma'_0 = -1.269874$	$\omega'_0 = +0.650478$
	$\sigma_0 = -0.623801$	$\omega_0 = -0.319535$
(Zeros)	$\sigma_{z0} = +0.0$	$\omega_{z0} = +1.082392$
(Zeros)	$\sigma_{z1} = +0.0$	$\omega_{z1} = +2.613126$

And, finally, we generate the transfer function from Equations 2.48 – 2.54. (Refer to Example 2.6 for an explanation of the two transfer functions.)

$$H^*_{I,4}(S) = \frac{0.0223872 \cdot (S^2 + 1.171573) \cdot (S^2 + 6.828427)}{(S^2 + 0.383548 \cdot S + 0.364590) \cdot (S^2 + 1.247603 \cdot S + 0.491231)}$$

$$H_{I,4}(S) = \frac{0.0223872 \cdot (S^2 + 4.686292) \cdot (S^2 + 27.313709)}{(S^2 + 0.767095 \cdot S + 1.458359) \cdot (S^2 + 2.495206 \cdot S + 1.964923)}$$

The results of using FILTER to determine these coefficients are presented in Figures 2.11 and 2.12 while the magnitude and phase responses are presented in Figures 2.13 and 2.14. The same procedure was used as in the Butterworth and Chebyshev cases.

2.5 Elliptic Normalized Approximation Functions

The elliptic or Cauer approximation function provides the best selectivity characteristic of any of the approximation methods discussed thus far. No other approximation method will be able to provide a lower-order filter for the specifications provided. The elliptic filter combines ripple in the passband and stopband in order to accomplish this feat. However, the elliptic approximation is also the most difficult to design. It involves the most sophisticated mathematical functions of any of the methods discussed in this text. Luckily, many good minds have laid the foundation for our work and their results will be presented here so that we can put the design procedure into a workable algorithm.

2.5.1 Elliptic Approximation Magnitude Response

The elliptic approximation's magnitude frequency response function is shown in Equation 2.55 where R_n is the Chebyshev rational function of order n. R_n is composed of both numerator and denominator portions which allow an equiripple response in both the passband and stopband.

$$\left|H_{E,n}[j(\omega / \omega_o)]\right| = \frac{1}{\sqrt{1 + \varepsilon^2 \cdot R_n^2(\omega_o / \omega)}} \tag{2.55}$$

where ε is as defined previously in Equation 2.56.

$$\varepsilon = \sqrt{10^{-0.1 \cdot a_{\text{pass}}} - 1} \tag{2.56}$$

```
                    Inv. Chebyshev 3rd Order Lowpass

                Filter implementation is:     Analog
                Filter approximation is:      Inv Cheby
                Filter selectivity is:        Lowpass
                Passband gain        (dB):    -1.00
                Stopband gain        (dB):    -22.00
                Passband frequency   (Hz):    0.16
                Stopband frequency   (Hz):    0.32

                              Order = 3
                       Overall Gain = 3.09341803e-001

      Numerator Coefficients                Denominator Coefficients
  [S^2 +       S          +      ()      ]  [S^2 +       S          +      ()      ]
  ==================================  ==================================
1 0.0 0.00000000e+00 1.54556433e+000  0.0 1.00000000e+00 1.54556433e+000
2 1.0 0.00000000e+00 5.33333333e+000  1.0 1.06745667e+00 1.64982295e+000
```

Figure 2.11 Inverse Chebyshev normalized third-order coefficients.

```
                    Inv. Chebyshev 4th Order Lowpass

                Filter implementation is:     Analog
                Filter approximation is:      Inv Cheby
                Filter selectivity is:        Lowpass
                Passband gain        (dB):    -1.00
                Stopband gain        (dB):    -33.00
                Passband frequency   (Hz):    0.16
                Stopband frequency   (Hz):    0.32

                              Order = 4
                       Overall Gain = 2.23872114e-002

      Numerator Coefficients                Denominator Coefficients
  [S^2 +       S          +      ()      ]  [S^2 +       S          +      ()      ]
  ==================================  ==================================
1 1.0 0.00000000e+00 4.68629150e+000  1.0 7.67095479e-01 1.45835865e+000
2 1.0 0.00000000e+00 2.73137085e+001  1.0 2.49520594e+00 1.96492342e+000
```

Figure 2.12 Inverse Chebyshev normalized fourth-order coefficients.

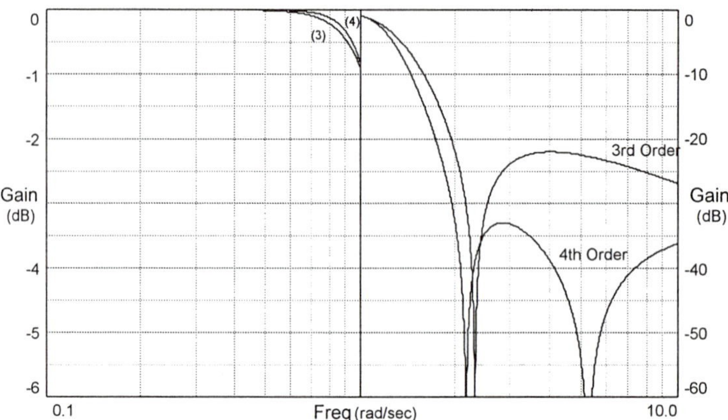

Figure 2.13 Inverse Chebyshev third-order and fourth-order magnitude responses.

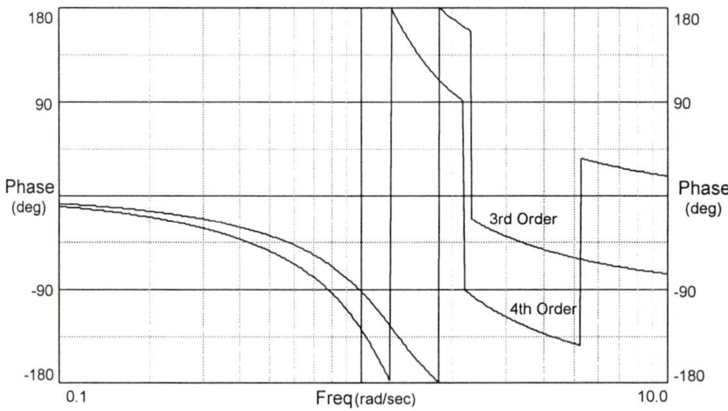

Figure 2.14 Inverse Chebyshev third-order and fourth-order phase responses.

The Chebyshev rational function R_n and much of elliptic approximation theory is based on the elliptic integral and the Jacobian elliptic functions. These functions can be evaluated via advanced mathematical packages available for most computers and we will discuss implementing them in C code later in this chapter. The incomplete elliptic integral of the first kind is shown in Equation 2.57, where k is referred to as the modulus and ϕ is the amplitude of the integral. The modulus k must be less than or equal to 1 for the elliptic integral to be real. The elliptic sine, cosine, tangent, and difference functions which are based on the elliptic integral are given in Equations 2.58 – 2.61, respectively.

$$u(\phi, k) = \int_0^{\phi} (1 - k^2 \sin^2 x)^{-1/2} dx \tag{2.57}$$

$$sn(u, k) = \sin(\phi) \tag{2.58}$$

$$cn(u, k) = \cos(\phi) \tag{2.59}$$

$$sc(u, k) = \tan(\phi) \tag{2.60}$$

$$dn(u, k) = \frac{d\phi}{du} \tag{2.61}$$

The complete elliptic integral of the first kind will be used more often than the incomplete integral and it is defined in Equation 2.62. It should be noted at this point that there are various ways to define the elliptic integrals and elliptic functions. Some use the modulus k as we have in this text, while others use other parameters related to k.

$$CEI(k) = u(\pi / 2, k) = \int_0^{\pi/2} (1 - k^2 \sin^2 x)^{-1/2} dx \tag{2.62}$$

2.5.2 Elliptic Approximation Order

The order of the elliptic approximation function required to meet the specifications for a filter is given in Equation 2.63.

$$n_E = \frac{CEI(rt) \cdot CEI(\sqrt{1 - kn^2})}{CEI(\sqrt{1 - rt^2}) \cdot CEI(kn)} \tag{2.63}$$

where CEI refers to the complete elliptic integral, and the ratio rt and the kernel kn are defined as

$$rt = \omega_{pass} / \omega_{stop} \tag{2.64}$$

$$kn = \sqrt{10^{-0.1a_{\text{pass}}} - 1) / (10^{-0.1a_{\text{stop}}} - 1)} \tag{2.65}$$

Example 2.8 Elliptic Order Calculation

Problem: Determine the order of an elliptic filter required to satisfy the following specifications:

$a_{\text{pass}} = -1$ dB, $a_{\text{stop}} = -34$ dB, $\omega_{\text{pass}} = 1$ rad/sec, and $\omega_{\text{stop}} = 2$ rad/sec

Solution: In order to determine the order of the elliptic approximation, we first determine that $rt = 0.5$ and $kn = 0.0101548$. Then using any appropriate math package we can determine that

$$n_E = \frac{1.686 \cdot 5.976}{2.157 \cdot 1.571} = 2.97$$

which indicates that a third-order filter will be required. Notice that the standard and inverse Chebyshev approximations require a fourth-order function to provide $a_{\text{stop}} = -33$ dB and a Butterworth approximation would require a seventh-order to meet this specification.

2.5.3 Elliptic Approximation Pole-Zero Locations

The pole and zero locations for the elliptic approximation function are also dependent on the elliptic integral and the elliptic functions defined in the previous section. We'll start by defining a variable v_o, which is used in the calculation of the pole and zero locations.

$$v_o = \frac{CEI(rt) \cdot sc^{-1}(\varepsilon^{-1}, kn)}{n \cdot CEI(kn)} \tag{2.66}$$

Next, the pole's real and imaginary components are determined as

$$\sigma_m = -\frac{cn[f(m), rt] \cdot dn[f(m), rt] \cdot sn(v_o, \sqrt{1 - rt^2}) \cdot cn(v_o, \sqrt{1 - rt^2})}{1 - dn^2[f(m), rt] \cdot sn^2(v_o, \sqrt{1 - rt^2})} \tag{2.67}$$

$$\omega_m = \frac{sn[f(m), rt] \cdot dn(v_o, \sqrt{1-rt^2})}{1 - dn^2[f(m), rt] \cdot sn^2(v_o, \sqrt{1-rt^2})} \tag{2.68}$$

where

$$f(m) = \frac{CEI(rt) \cdot (2 \cdot m + 1)}{n}, \quad m = 0,\ 1,\ \ldots\ ,\ (n/2) - 1\ \ (n \text{ even}) \tag{2.69a}$$

$$f(m) = \frac{CEI(rt) \cdot (2 \cdot m + 2)}{n}, \quad m = 0,\ 1,\ \ldots\ ,\ [(n-1)/2] - 1\ \ (n \text{ odd}) \tag{2.69b}$$

Note the negative sign for σ_m, which effectively moves the pole location from the first quadrant to the second quadrant.

In the case of odd-order approximations, the first-order denominator pole will be located on the negative real axis at

$$\sigma_R = -\frac{sn(v_o, \sqrt{1-rt^2}) \cdot cn(v_o, \sqrt{1-rt^2})}{1 - sn^2(v_o, \sqrt{1-rt^2})} \tag{2.70}$$

And finally the location of the zeros which will be purely imaginary on the $j\omega$ axis are given by

$$\sigma_{zm} = 0.0 \tag{2.71}$$

$$\omega_{zm} = \frac{1}{rt \cdot sn[f(m), rt]} \tag{2.72}$$

Although the elliptic approximation requires a number of mathematical functions which aren't in everyday usage, we have most of the hard work done in determining the transfer function we need. Our primary objective in this section is to develop an orderly manner to calculate the pole and zero locations.

2.5.4 Elliptic Approximation Transfer Functions

Now we are able to define the first-order and quadratic factors that will make up the elliptic approximation function. The first-order factor for the elliptic approximation is indicated in Equation 2.73, where σ_R is as indicated in Equation 2.70. Again, there is no matching finite zero for the first-order pole factor; it is located at infinity.

$$H_o(S) = \frac{\sigma_R}{S + \sigma_R} \tag{2.73}$$

The form of the quadratic components of the transfer function will also be identical to the inverse Chebyshev case as indicated below.

$$H_m(S) = \frac{B_{2m} \cdot (S^2 + A_{1m} \cdot S + A_{2m})}{A_{2m} \cdot (S^2 + B_{1m} \cdot S + B_{2m})} \tag{2.74}$$

where

$$B_{1m} = -2 \cdot \sigma_m \tag{2.75}$$

$$B_{2m} = \sigma_m^2 + \omega_m^2 \tag{2.76}$$

$$A_{1m} = -2 \cdot \sigma_{zm} = 0.0 \tag{2.77}$$

$$A_{2m} = \sigma_{zm}^2 + \omega_{zm}^2 = \omega_{zm}^2 \tag{2.78}$$

We are now ready to define a generalized transfer function for the elliptic approximation function which is almost identical to the inverse Chebyshev case. The difference lies in the ripple in the passband as in the standard Chebyshev case. Consequently, the even-order ripple adjustment factor is included in Equation 2.79. The ratio of product factors is combined with this value to determine the total gain adjustment.

$$H_{E,n}(S) = \frac{(10^{0.05 \cdot a_{\text{pass}}}) \cdot \prod_m (B_{2m}) \cdot \prod_m (S^2 + A_{1m} \cdot S + A_{2m})}{\prod_m (A_{2m}) \cdot \prod_m (S^2 + B_{1m} \cdot S + B_{2m})}, \tag{2.79a}$$

$m = 0,\ 1,\ \text{K}\ ,\ (n/2) - 1\ \ (n \text{ even})$

$$H_{E,n}(S) = \frac{\sigma_R \cdot \prod_m (B_{2m}) \cdot \prod_m (S^2 + A_{1m} \cdot S + A_{2m})}{(S + \sigma_R) \cdot \prod_m (A_{2m}) \cdot \prod_m (S^2 + B_{1m} \cdot S + B_{2m})}, \tag{2.79b}$$

$m = 0,\ 1,\ \text{K}\ ,\ [(n-1)/2] - 1\ \ (n \text{ odd})$

Numerical examples of the determination of elliptic filter coefficients are now in order.

Example 2.9 Elliptic Third-Order Normalized Transfer Function

Problem: Determine the order, pole and zero locations, and transfer function coefficients for an elliptic filter to satisfy the following specifications:

$a_{\text{pass}} = -1$ dB, $a_{\text{stop}} = -34$ dB, $\omega_{\text{pass}} = 1$ rad/sec, and $\omega_{\text{stop}} = 2$ rad/sec

Solution: First, we determine the fundamental constants needed from Equations 2.56 and 2.63 – 2.66.

$\varepsilon = 0.508847$
$rt = 0.50$
$CEI(rt) = 1.685750$
$CEI[\text{sqrt}(1 - rt^2)] = 2.156516$
$v_o = 0.510786$

$n = 2.97$ (3rd order)
$kn = 0.0101549$
$CEI(kn) = 1.570837$
$CEI[\text{sqrt}(1 - kn^2)] = 5.976226$

Next, we find the locations of the first-order pole, the complex pole in the second quadrant, and the second-order zeros on the $j\omega$ axis from Equations 2.67 – 2.72.

(1st order)	$\sigma_R = -0.539953$	$\omega_R = 0.0$
$f(0) = 1.123834$	$\sigma_0 = -0.217032$	$\omega_0 = +0.981574$
(Zeros)	$\sigma_{z0} = +0.0$	$\omega_{z0} = +2.270068$

And, finally, we generate the transfer function from Equations 2.73 – 2.79.

$$H_{E,3}(s) = \frac{0.539954 \cdot 1.010590 \cdot (S^2 + 5.153209)}{5.153209 \cdot (S + 0.539954) \cdot (S^2 + 0.434064 \cdot S + 1.010590)}$$

Example 2.10 Elliptic Fourth-Order Normalized Transfer Function

Problem: Determine the order, pole and zero locations, and transfer function coefficients for an elliptic filter to satisfy the following specifications:

$a_{pass} = -1$ dB, $a_{stop} = -51$ dB, $\omega_{pass} = 1$ rad/sec, and $\omega_{stop} = 2$ rad/sec

Solution: First, we determine the fundamental constants needed from Equations 2.56 and 2.63 – 2.66.

$\varepsilon = 0.508847$	$n = 3.95$ (4th order)
$rt = 0.50$	$kn = 0.00143413$
$CEI(rt) = 1.685750$	$CEI(kn) = 1.570797$
$CEI[\text{sqrt}(1 - rt^2)] = 2.156516$	$CEI[\text{sqrt}(1 - kn^2)] = 7.933494$
$v_o = 0.383119$	

Next, we find the locations of the two complex poles in the second quadrant and the second-order zeros on the jω axis from Equations 2.67 – 2.72.

$f(0) = 0.421438$	$\sigma_0 = -0.351273$	$\omega_0 = +0.442498$
$f(1) = 1.264313$	$\sigma_1 = -0.121478$	$\omega_1 = +0.989176$
(Zeros)	$\sigma_{z0} = +0.0$	$\omega_{z0} = +4.922113$
(Zeros)	$\sigma_{z0} = +0.0$	$\omega_{z0} = +2.143189$

And, finally, we generate the transfer function from Equations 2.73 – 2.79. Note that in this even-order case, the gain constant of 0.891251 is included.

$$H_{E,4}(S) = \frac{0.00253912 \cdot (S^2 + 24.227201) \cdot (S^2 + 4.593261)}{(S^2 + 0.702546 \cdot S + 0.319197) \cdot (S^2 + 0.242957 \cdot S + 0.993226)}$$

The results from using FILTER to determine the coefficients are given in Figures 2.15 and 2.16, while the magnitude and phase responses for the elliptic filters are presented in Figures 2.17 and 2.18.

2.6 Comparison of Approximation Methods

Now that we have discussed the four approximation methods and displayed third-order and fourth-order magnitude and phase plots, we are in

```
                    Elliptic 3rd Order Lowpass

             Filter implementation is:      Analog
             Filter approximation is:       Elliptic
             Filter selectivity is:         Lowpass
             Passband gain        (dB):     -1.00
             Stopband gain        (dB):     -34.00
             Passband frequency   (Hz):     0.16
             Stopband frequency   (Hz):     0.32

                          Order = 3
                   Overall Gain = 1.96108843e-001

      Numerator Coefficients             Denominator Coefficients
  [S^2 +       S         +       ()     ]  [S^2 +       S        +       ()     ]
  ==================================  ==================================
1 0.0 0.00000000e+00 5.39953774e-001  0.0 1.00000000e+00 5.39953774e-001
2 1.0 0.00000000e+00 5.15320912e+000  1.0 4.34064064e-01 1.01058988e+000
```

Figure 2.15 Elliptic normalized third-order coefficients.

```
                    Elliptic 4th Order Lowpass

             Filter implementation is:      Analog
             Filter approximation is:       Elliptic
             Filter selectivity is:         Lowpass
             Passband gain        (dB):     -1.00
             Stopband gain        (dB):     -51.00
             Passband frequency   (Hz):     0.16
             Stopband frequency   (Hz):     0.32

                          Order = 4
                   Overall Gain = 2.53911537e-003

      Numerator Coefficients             Denominator Coefficients
  [S^2 +       S         +       ()     ]  [S^2 +       S        +       ()     ]
  ==================================  ==================================
1 1.0 0.00000000e+00 2.42272012e+001  1.0 7.02545661e-01 3.19196826e-001
2 1.0 0.00000000e+00 4.59326053e+000  1.0 2.42956738e-01 9.93226262e-001
```

Figure 2.16 Elliptic normalized fourth-order coefficients.

a position to compare the results. First, we look at the magnitude plots of Figures 2.5, 2.9, 2.13, and 2.17. Table 2.1 shows the gains achieved at the stopband edge frequency of 2 rad/sec for each normalized filter type and order. (Each filter was designed with a passband gain of –1 dB.) Obviously, if

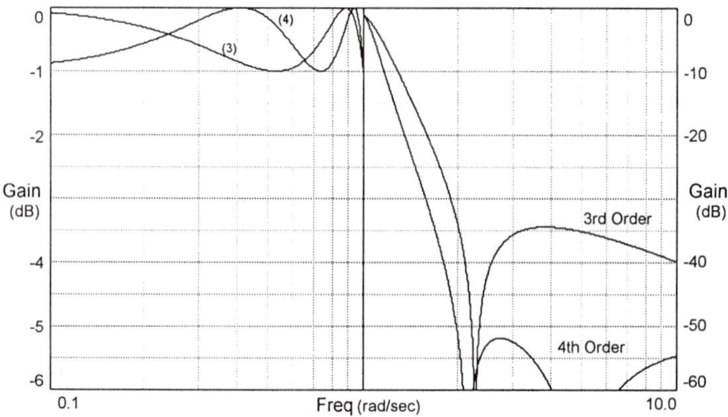

Figure 2.17 Elliptic third-order and fourth-order magnitude responses.

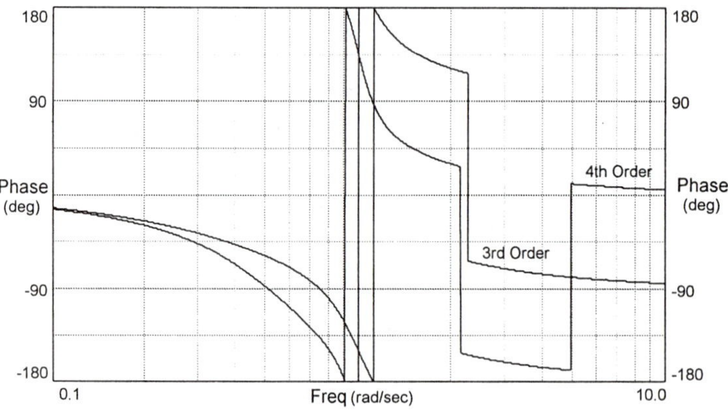

Figure 2.18 Elliptic third-order and fourth-order phase responses.

attenuation characteristics in the stopband are the primary concern, an elliptic filter would have to be the choice. It provides 12 dB more attenuation than the Chebyshev types and 22 dB more attenuation than the Butterworth filter for the third-order case. In the fourth-order case, the differences increase to over 18 and 33 dB compared to the Chebyshev and Butterworth filters. The Chebyshev filter types themselves afford better stopband characteristics when compared to the Butterworth filter. They provide 10 and 15 dB more attenuation for the third-order and fourth-order cases. Although the table only list the gains for third-order and fourth-order filters, the same trend continues for higher-order filters.

Table 2.1 Comparison of filter gains at 2 rad/sec

	3rd Order	4th Order
Butterworth	–12.5 dB	–18.3 dB
Chebyshev	–22.5 dB	–33.8 dB
Inv. Cheby.	–22.5 dB	–33.8 dB
Elliptic	–34.5 dB	–51.9 dB

Although the Chebyshev and inverse Chebyshev filters provide the same gains at the passband and stopband edge frequencies, their responses are not identical. If we were to take a close look at the frequency response in the passband, we would find that the inverse Chebyshev provides a better approximation to the ideal response except at frequencies very near to 1 (the normalized passband edge frequency). In that case, the standard Chebyshev produces a tighter fit. In the transition band, the standard Chebyshev response provides a more rapid transition. And in the stopband the standard Chebyshev's response continues to increase the attenuation as the frequency increases, while the inverse Chebyshev's response alternates between small gains and a_{stop}. In some cases, the filter designer might trade the faster transition for the nondecreasing attenuation.

Although the magnitude characteristics of a filter are very important, the phase characteristics of a filter also are crucial in many projects. Whether it be audio networks or data transmission systems, designers are looking for filters with linear phase response. Nonlinear phase response in an audio network will cause noticeable phase distortion for the listener which cannot be tolerated, especially in high-quality systems. In data transmission systems nonlinear phase response produces group delays which are functions of frequency. This produces distortion in the pulses sent over the system and can distort edges and levels to the point of causing errors in the received signal. We can compare the phase responses of Figures 2.6, 2.10, 2.14, and 2.18 to see the level of phase distortion for each approximation type. Remember that the transitions from –180 to +180 degrees are not discontinuities, but rather a function of the display method. (The phase response is written to a data file in its true form.) Table 2.2 shows the phase angles for the third-order and fourth-order filters at the passband and stopband edge frequencies.

Table 2.2 Comparison of filter phase at 1 and 2 rad/sec`

	3rd @ 1 r/s	3rd @ 2 r/s	4th @ 1 r/s	4th @ 2 r/s
Butterworth	–104°	–192°	–146°	–266°
Chebyshev	–154°	–238°	–230°	–330°
Inv. Cheby.	-94°	–192°	–133°	–264°
Elliptic	–150°	–238°	–226°	-330°

As Table 2.2 and the phase plots indicate, the filters with the maximally flat response in the passband (Butterworth and inverse Chebyshev) provide the most linear response, although the inverse Chebyshev does have phase discontinuities in the stopband caused by the complex zeros. These are usually not critical because the filter's magnitude response is very small at these frequencies and the distortion should be minimal. The phase responses of the standard Chebyshev and elliptic are also matched very closely and can be judged equivalent except for the discontinuities in the stopband caused by the zeros for the elliptic case.

A filter designer's task is not always clear cut. It seems that every project requires as much stopband attenuation as possible while providing a phase response as linear as possible. The task becomes one of weighing the importance of each characteristic. If phase response is more critical than magnitude response, then the Butterworth filter is a better choice. If the opposite is true, the elliptic filter is a better choice. If magnitude and phase responses are nearly equal in importance, than one of the Chebyshev filters may be the best choice. Other alternatives are also possible. Elliptic filters can be used for their selectivity, with phase compensation filters added to make the phase more linear. (These filter types are not covered in this text, but references in the analog filter design section of Appendix A can provide further information.) A designer must be careful when pursuing these alternatives since in some cases the result may be no better than the equivalent Butterworth or Chebyshev filter.

2.7 C Code for Normalized Approximation Functions

Now that we have determined approximation functions for the four most popular methods used in analog filter design, we are ready to implement them using C code. In Chapter 1, we developed an outline for our filter design project which is indicated by the Chapter 1 version below. However, we have now reached the point in our spiraling technique where the analog design item listed under I.B.1 needs further definition (as shown in the expanded version).

Analog and Digital Filter Design Project – Chapter 1

- I. Design analog or digital filter.
  - A. Get filter specifications from user.
    1. Get implementation type (analog/dig IIR/dig FIR).
    2. Get selectivity type (LP/HP/BP/BR).
    3. Get approximation type (Buttr/Chby/IChby/Ellip).
    4. Get passband and stopband edge frequencies.
    5. Get passband and stopband gains.
    6. Get a descriptive title for project.
  - B. Design filter according to specifications.
    1. Design analog, if designated.
    2. Design digital FIR, if designated.
    3. Design digital IIR, if designated.
  - C. Display filter parameters on screen.
    1. Display filter input specifications.
    2. Display filter design values.
  - D. Save filter parameters to file.
    1. Save parameters in text form.
    2. Save parameters in binary form.
- II. Determine frequency response of filter.
- III. Display magnitude and phase response on screen.

Analog and Digital Filter Design Project – Expanded Version

- I. Design analog or digital filter.
  - A. Get filter specifications from user.
  - B. Design filter according to specifications.
    1. Design analog, if designated.
       - a. Calculate required order.
       - b. Calculate coefficients.
         - (1) Butterworth.
         - (2) Chebyshev.
         - (3) Inverse Chebyshev.
         - (4) Elliptic.
       - c. Unnormalize coefficients.
    2. Design digital FIR, if designated.
    3. Design digital IIR, if designated.
  - C. Display filter parameters on screen.
  - D. Save filter parameters to file.
- II. Determine frequency response of filter.
- III. Display magnitude and phase response on screen.

In this section, we will discuss the C functions necessary to implement all of the requirements under items I.B.1.a and I.B.1.b.

2.7.1 Calc_Filter_Coefs Function and Error Handling

In the main program we developed in Chapter 1, we called `Calc_Filter_Coefs` in order to calculate the necessary filter coefficients for the user specified design. This function represents our implementation of item I.B from our project outline above. We can now see how that function is put together in Listing 2.1. As we can see, the primary purpose for this function is to use the filter's implementation type as a way of determining which of three other functions to call. `Calc_DigFIR_Coefs` and `Calc_DigIIR_Coefs` will be discussed in Chapters 7 and 8, respectively, and `Calc_Analog_Coefs` will be discussed in the next section. The user's specifications are transferred to the next appropriate function via `FP`, the pointer to the `Filt_Params` structure. This function allows us to add other methods of filter design to the project at a later time. The only changes to our project would be the addition of more choices to this function, and the necessary code to support those choices.

If any one of the calculation functions generates an error, the error number is processed in a special way in order to provide the maximum amount of information to the user. There are a number of methods which could be used to handle the disruption of the normal program execution, and the best method depends on the size and type of project being handled. (There are several excellent references given at the end of the text for further information on this matter.) Our project is rather small in the scheme of software development, and will not have too many layers of function calls. Most of the errors in this project are expected to occur in the development phase, and failure of this program will usually not cost the user much in data or time. Using this characterization of the project, the following error handling method is justified.

Global variables are one way to handle error conditions in a complex program, but will not be used in this project since they are "inelegant" and prone to misuse. Instead we will develop a method which propagates the error condition up the call chain while keeping unique information about the exact path of the propagation. This allows the `main` program to handle all error conditions while also providing the necessary information to the user or programmer. The technique which we will use is illustrated in Figure 2.19. Assuming that an error has occurred in the function Func3, we follow the error code as it is returned to the `main` function. The error code (3) returned to Func2 is first multiplied by 10 to move it to the 10's digit, then an appropriate error code is added to it (in this case 2). Then the error code 32 is returned to Func1 where it is again multiplied by 10 before Func1 adds its error value. Finally, the error code 321 is returned to the `main` function and can be displayed for the user or programmer. The error number contains important information. The 3 refers to the error number returned at the lowest level, while the 2 is the error number returned by the middle level, and the 1 is the error from the function closest to main. This information can lead us directly to the offending function and, it is hoped, to an area within the function. This method does have some potential problems. For instance, if more than five levels of layering are involved in the project, the error

```
/*=======================================================
  Calc_Filter_Coefs() - determines implementation and
  calls appropriate calculation function
  Prototype:  int Calc_Filter_Coefs(Filt_Params *FP);
  Return:     error value
  Arguments:  FP - ptr to struct holding filter params
=======================================================*/
int Calc_Filter_Coefs(Filt_Params *FP)
{ int Error;                      /*  error value */

  /*  Call correct calc function for analog,
      digital FIR or IIR filters. */
  switch(FP->implem)
  { case 'A':
      Error = Calc_Analog_Coefs(FP);
      if(Error) { return 10*Error+1;}
      break;
    case 'F':
      Error = Calc_DigFIR_Coefs(FP);
      if(Error) { return 10*Error+2;}
      break;
    case 'I':
      Error = Calc_DigIIR_Coefs(FP);
      if(Error) { return 10*Error+3;}
      break;
    default:
      return ERR_FILTER;
  }
  return ERR_NONE;
}
```

Listing 2.1 Calc_FIlter_Coefs function.

variable would have to be stored as a `long` or 32-bit integer. And if more than 10 different errors could occur in a function, the error number would take up two digits in the error code, and this would reduce the number of layers which could be handled. This method is primarily structured for the developer/user as stated earlier, and if a project was being developed for the more general user, more effort would have to go into the error reporting technique.

2.7.2 Calc_Analog_Coefs Function

The `Calc_Analog_Coefs` function is given in Listing 2.2 and represents item I.B.1 in our project outline. It is designed primarily to organize the calculation process, not to perform the calculations explicitly. The first function called is `Calc_Filter_Order`, which determines the filter order from the user specifications in `FP`. Next the normalized filter coefficients are determined using `Calc_Normal_Coefs`, which stores the coefficients in the `Filt_Params` structure. And, finally, the coefficients are unnormalized to lowpass, highpass, bandpass, or bandstop coefficients at the user specified frequencies via the `Unnormalize_Coefs` function. These functions, respectively, represent item I.B.1.a-c in our project outline and can be found in the \FILTER\F_DESIGN.C module.

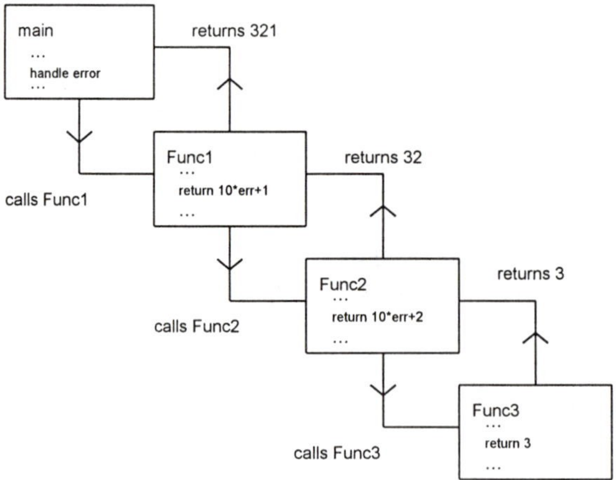

Figure 2.19 Error handling method.

```
/*===========================================================
  Calc_Analog_Coefs() - calcs normal analog coefs
  Prototype:  int Calc_Analog_Coefs(Filt_Params *FP);
  Return:     error value
  Arguments:  FP - ptr to struct holding filter params
===========================================================*/
int Calc_Analog_Coefs(Filt_Params *FP)
{ int Error;                  /*  error value */

  /*  Determine filter order, then normal coefs,
      then unnormalize them. */
  Error = Calc_Filter_Order(FP);
  if(Error) { return 10*Error+1;}
  Error = Calc_Normal_Coefs(FP);
  if(Error) { return 10*Error+2;}
  Error = Unnormalize_Coefs(FP);
  if(Error) { return 10*Error+3;}
  return ERR_NONE;
}
```

Listing 2.2 Calc_Analog_Coefs function.

2.7.3 Calc_Normal_Coefs Function

In the case of the `Calc_Normal_Coefs` function we are implementing item I.B.1.b of our project outline. In this function, as shown in Listing 2.3, we must allow for individual functions to carry out the actual coefficient calculation. However, before we actually make the coefficient calculations, we allocate memory for the storage of the coefficients. By taking care of this necessary task in `Calc_Normal_Coefs`, we eliminate the need to handle it in each of the four individual functions. It is important to remember that by simply reserving a place in the `Filt_Params` structure for the pointers `acoefs` and `bcoefs`, we have *not* reserved any memory for the coefficients themselves. The pointers were originally set to zero or NULL by the `calloc` command we used to allocate memory for the `Filt_Params` structure. This effectively says that there is no memory available for coefficient storage, and if we try to access a coefficient while in this state, we would get an error. In fact, we hope we get an error to let us know that there is a problem in our algorithm. It would be far worse to have the pointers hold a nonzero value which is pointing to some random address in memory. In that case, we would get no error, but the values we would be accessing would be random nonsense.

```
/*=====================================================
  Calc_Normal_Coefs() - allocates memory for coefs and
              calls proper function to calc coefs
  Prototype:  int Calc_Normal_Coefs(Filt_Params *FP);
  Return:     error value
  Arguments:  FP - ptr to struct holding filter params
=====================================================*/
int Calc_Normal_Coefs(Filt_Params *FP)
{ int Number_Coefs,    /*  Number of coefs in array */
      Error;           /*  error value */

  /*  Allocate memory for coefs.  There are 3 coefs
      for each quadratic. First-order factors are
      considered as quadratics. */
  Number_Coefs = 3 * ((FP->order + 1) / 2);
  FP->acoefs = (double *)
               malloc(Number_Coefs * sizeof(double));
  if(!FP->acoefs) { return ERR_ALLOC;}
  FP->bcoefs = (double *)
               malloc(Number_Coefs * sizeof(double));
  if(!FP->bcoefs) { return ERR_ALLOC;}
  /*  Calculate coefs based on approximation. */
  switch (FP->approx)
  { case 'B': Error = Calc_Butter_Coefs(FP);
              if(Error) { return 10*Error+1;}
              break;
    case 'C': Error = Calc_Cheby_Coefs(FP);
              if(Error) { return 10*Error+2;}
              break;
    case 'E': Error = Calc_Ellipt_Coefs(FP);
              if(Error) { return 10*Error+3;}
              break;
    case 'I': Error = Calc_ICheby_Coefs(FP);
              if(Error) { return 10*Error+4;}
              break;
    default:  return ERR_FILTER;
  }
  return ERR_NONE;
}
```

Listing 2.3 Calc_Normal_Coefs function.

Up to this point in the program, we did not have the necessary information to determine the number of coefficients in each of the acoefs and `bcoefs` arrays. But since we now have the order of the filter, we can determine the number of coefficients exactly in the following manner. Each quadratic factor of the approximation function will require three coefficients (the s^2-term coefficient, the s-term coefficient, and the constant term coefficient). In addition, we will treat any odd-order approximation first-order factor as a quadratic. We can make use of the integer math in C to simplify the calculation for the number of coefficients based on the order of the filter as shown in Equation 2.80. Any time two integers are divided and the result stored in another integer, the result of the division is truncated prior to assignment. If a sixth-order filter is being processed, the internal calculation of (6 + 1)/2 would result in a value of 3 (the number of quadratics), and therefore the number of coefficients would be calculated as nine. A fifth-order filter will also require three quadratics (nine coefficients) since the first-order factor will be stored as a quadratic.

$$\text{Number\_Coef}s = 3 \cdot [(\text{FP}->\text{order} + 1) / 2] \tag{2.80}$$

Once the correct number of coefficients has been determined, `malloc` is used to allocate memory for that number of `doubles`, and if an error occurs, we leave the function. The `malloc` function is used in place of `calloc` because the `calloc` function takes a bit longer to execute since it fills the memory with zeros. We will be writing right over the numbers in memory anyway so there is no reason to waste time filling them with zeros first.

The coefficient arrays are organized in the following manner. The `acoefs` array stores the coefficients for the numerator factors while the `bcoefs` array will store the coefficients for the denominator factors. (`acoefs` and `bcoefs` will take on different meanings in the digital filter design so for now we can remember A for above the line and B for below the line.) If the approximation function has an odd-order, the first-order coefficients are stored as a quadratic in the first three coefficients of each array with the s^2-term set to zero. The next three coefficients are for the first true quadratic, and then all other quadratics follow. If the approximation function has an even-order, then the first three coefficients are for the first quadratic, the next three coefficients for the second quadratic, and so on. Within the three coefficients, the first coefficient always refers to the s^2-term, the second refers to the s-term coefficients, and the last coefficient is the constant term in the quadratic factor.

The individual functions for calculating the normalized coefficients for the four approximation methods can be found on the software disk in the

\FILTER\F_DESIGN.C module. They all determine the coefficients in a manner consistent with our development earlier in this chapter.

2.7.4 Advanced Math Functions

There are a number of advanced math functions required by our approximation functions. These functions are *not* contained in F_DESIGN.C as are the rest of the functions discussed in this chapter, but are instead a part of the \FILTER\ADV_MATH.C module. These include the asinh and acosh functions used in the Chebyshev functions as well as the elliptic integral and Jacobian elliptic functions necessary to define the elliptic approximation. The values of these functions are typically calculated by the arithmetic-geometric mean method of iteration. A discussion of this method and the elliptic functions in general is beyond the scope of this text, but references have been provided in the analog and digital filter design sections of Appendix A.

The `Ellip_Integral` function is given in Listing 2.4 and uses the arithmetic-geometric mean method of determining the complete elliptic integral as defined in Equation 2.62. It takes as an argument the modulus k and returns the value of the complete elliptic integral. MAX_TERMS and ERR_SMALL have been defined as 100 and 1E-15, respectively, in the ADV_MATH.H include file. Of course these values could be changed as necessary.

```
/*==========================================================
  Ellip_Integral() - calcs complete elliptic integral
                     using arithmetic-geometric mean method
  Prototype:  void Ellip_Integral(double k);
  Return:     complete elliptic integral value
  Arguments:  k - the modulus of the integral
==========================================================*/
double Ellip_Integral(double k)
{ int      i;       /*  Loop counter. */
  double   A[MAX_TERMS],B[MAX_TERMS],
           C[MAX_TERMS]; /* Array storage values. */

  /*  Square the modulus as required by this method.*/
  k = k * k;
  /*  Initialize the starting values. */
  A[0] = 1;
  B[0] = sqrt(1-k);
  C[0] = sqrt(k);
```

```
    /*  Iterate until error is small enough. */
    for(i = 1; i < MAX_TERMS ;i++)
    { A[i] = (A[i-1] + B[i-1])/2;
      B[i] = sqrt(A[i-1]*B[i-1]);
      C[i] = (A[i-1] - B[i-1])/2;
      if(C[i] < ERR_SMALL)
      { break;}
    }
    return PI / (2 * A[i]);
}
```

Listing 2.4 Ellip_Integral function.

The `Ellip_Funcs` function as shown in Listing 2.5 calculates the Jacobian elliptic sine, cosine, and difference functions within the same function. The addresses of the three values are sent as arguments of the function, and when the values have been determined, the results are placed in the appropriate memory location. There are three key features of this technique to remember. First, when calling the function (as `Ellip_Funcs` is called from `Calc_Ellip_Coefs`), use the address of operator & to indicate that the argument to be sent is the *address of the variable* (for example `&SN`). Second, when defining the `Ellip_Funcs` function, indicate the argument using the dereferencing symbol * (as in `*sn`). And, finally, within the `Ellip_Funcs` function, whenever the value of the variable is referenced, use * in front of it to indicate that you want access to *where the variable is pointing*. For example, one of the last statements in the function reads `*sn = sin(P[0]);`. This simply instructs the machine to put the value of `sin(P[0])` in the place where sn is pointing.

```
/*=======================================================
  Ellip_Funcs() - calcs Jacobian elliptic functions
                  using arithmetic-geometric mean method
  Prototype:    void Ellip_Funcs(double u,double k,
                         double *sn,double *cn,double *dn);
  Return:       none. (values return via arg list)
  Arguments:    u - the value of the elliptic integral
                k - the modulus of the integral
                sn - Jacobian elliptic sine function
                cn - Jacobian elliptic cosine function
                dn - Jacobian elliptic differenc func
=======================================================*/
void Ellip_Funcs(double u,double k,double *sn,
                                double *cn,double *dn)
```

```
{ int     i,imax;    /*  Loop counter and max value. */
  double  A[MAX_TERMS],B[MAX_TERMS],/* Array storage*/
          C[MAX_TERMS],P[MAX_TERMS];/* values. */

  /* Square the modulus as required in this method.*/
  k = k * k;
  /* Initialize the starting values. */
  A[0] = 1;
  B[0] = sqrt(1-k);
  C[0] = sqrt(k);
  /*  Iterate until error is small enough. */
  for(i = 1; i < MAX_TERMS ;i++)
  { A[i] = (A[i-1] + B[i-1])/2;
    B[i] = sqrt(A[i-1]*B[i-1]);
    C[i] = (A[i-1] - B[i-1])/2;
    if(C[i] < ERR_SMALL)
    { break;}
  }
  /*  Get last value of i. */
  if(i == MAX_TERMS)
  { imax = i - 1;}
  else
  { imax = i;}
  /*  Determine phase and unwrap it. */
  P[imax] = pow(2,imax) * A[imax] * u;
  for(i = imax; i > 0 ; i--)
  { P[i-1] = (asin(C[i]*sin(P[i])/A[i]) + P[i]) / 2;}
  /*  Place values in memory locations specified. */
  *sn = sin(P[0]);
  *cn = cos(P[0]);
  *dn = sqrt(1 - k * (*sn) * (*sn));
}
```

Listing 2.5 Ellip_Funcs function.

2.8 Conclusion

In this chapter, we have studied the core of analog filter design, the normalized approximation functions. By developing these functions, we have laid the foundation for the remainder of the chapters on analog filter design as well as a good bit of digital IIR filter design. By approaching each approximation function in the same manner, and developing methods for determining exact pole and zero placement, we were able to generate the C code necessary to implement these algorithms in a clean, efficient manner. Our project outline has now reached the lowest level of implementation after spiraling through the project several times. In the next chapter, we will finish up the analog filter design calculations by determining a technique to unnormalize the transfer functions we have just developed.

Chapter 3

Analog Lowpass, Highpass, Bandpass, and Bandstop Filters

In the last chapter, we were able to determine the normalized approximation functions for the most common types of analog filters. Our task in this chapter is to unnormalize those approximation functions in a manner to produce lowpass, highpass, bandpass, and bandstop filters at the desired frequencies. This unnormalization will be carried out in such a way that the design of the normalized approximation functions will be central to the development. Figure 3.1 shows the three-step procedure used in the unnormalization. The simplicity of this procedure is the fact that the second step is the same for all filter design methods.

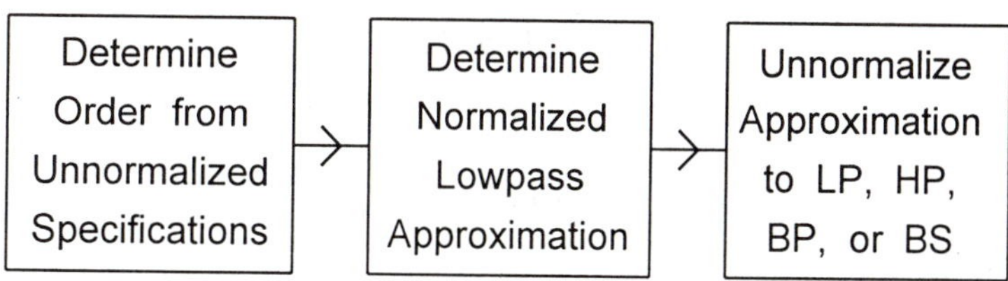

Figure 3.1 Procedure for unnormalization.

After determining the methods necessary for the unnormalization of the four types of filters, we will determine the C code necessary for this procedure. And, finally, we will generate the C code for displaying and saving the coefficients which we have determined.

3.1 Unnormalized Lowpass Approximation Functions

Even though the normalized approximation functions determined in the previous chapter are lowpass functions, they still need to be unnormalized to the proper operational frequency. The first step in the unnormalization procedure, as indicated in Figure 3.1, is to determine the order of the approximation function from the unnormalized specifications. The order of approximation function depends only on the passband and stopband gains and frequencies. The gains for both the normalized and unnormalized approximation functions will be the same, the only specifications which change are the passband and stopband edge frequencies. However, as indicated in Chapter 2, it is not the individual frequencies which determine the order of the approximation function, but rather the ratio of the frequencies. Therefore, we can define a frequency ratio variable in Equation 3.1 which will be used for the lowpass filter type as indicated by the additional subscript L. Note that it really makes no difference whether the frequencies are specified in radians per second or Hertz since the ratio will be the same.

$$\Omega_{rL} = \frac{\omega_{\text{stop}}}{\omega_{\text{pass}}} = \frac{f_{\text{stop}}}{f_{\text{pass}}} \tag{3.1}$$

Each of the equations from Chapter 2 which were used to determine the order of a particular filter type can now be redefined in terms of Ω_r as indicated in Equations 3.2 – 3.6.

$$n_B = \frac{\log[(10^{-0.1 \cdot a_{\text{stop}}} - 1)/(10^{-0.1 \cdot a_{\text{pass}}} - 1)]}{2 \cdot \log(\Omega_r)} \tag{3.2}$$

$$n_C = n_I = \frac{\cosh^{-1}\left[\sqrt{(10^{-0.1 \cdot a_{\text{stop}}} - 1)/(10^{-0.1 \cdot a_{\text{pass}}} - 1)}\right]}{\cosh^{-1}(\Omega_r)} \tag{3.3}$$

$$n_E = \frac{CEI(rt) \cdot CEI(\sqrt{1 - kn^2})}{CEI(\sqrt{1 - rt^2}) \cdot CEI(kn)} \tag{3.4}$$

where

$$rt = 1/\Omega_r \tag{3.5}$$

$$kn = \sqrt{(10^{-0.1a_{\text{pass}}} - 1)/(10^{-0.1a_{\text{stop}}} - 1)} \tag{3.6}$$

It may appear that we are going to a lot of work just to change a variable name, but Ω_r will be defined differently for each of the other types of filter selectivities as we will see in the next sections. For that reason, Equations 3.2 – 3.6 don't include the additional subscript L, however in each section we will define Ω_r with a subscript (as in Equation 3.1) for clarity.

So in this lowpass case, the first step in the unnormalization procedure doesn't require any work at all. We simply determine the order of the filter as we have in the past. The second step of the unnormalization procedure, the determination of the normalized approximation function, has already been developed in the previous chapter. It appears that we are ready to determine the third and final step of the procedure which is to unnormalize the normalized approximation function. In the lowpass case, this simply requires a scaling of the frequency characteristic from 1 rad/sec to a more usable frequency. A simple substitution for the normalized variable S is all that is necessary as shown in Equation 3.7. (A subscript of L is used to indicate that this substitution is for lowpass filters only.) The frequency constant ω_o will be ω_{stop} for the inverse Chebyshev approximation, as discussed in Chapter 2, and ω_{pass} for all other approximations.

$$S_L = \frac{s}{\omega_o} \tag{3.7}$$

3.1.1 Handling a First-Order Factor

We will be developing code to implement the unnormalization process, so it is important to carefully describe the substitution process. In the case of a first-order factor, the unnormalization process begins with Equation 3.8 where the B_1 coefficient is typically 1.

$$H(s) = \left.\frac{A_1 \cdot S + A_2}{B_1 \cdot S + B_2}\right|_{S=s/\omega_o} = \frac{A_1 \cdot (s/\omega_o) + A_2}{B_1 \cdot (s/\omega_o) + B_2} \tag{3.8}$$

In this equation, the uppercase A's and B's represent the coefficients of the normalized approximation function. After simplification Equation 3.9 results in a new set of coefficients. In this equation, the lowercase a's and b's represent the unnormalized coefficients which will be used in our final approximation function.

$$H(s) = \frac{A_1 \cdot s + A_2 \cdot \omega_o}{B_1 \cdot s + B_2 \cdot \omega_o} = \frac{a_1 \cdot s + a_2}{b_1 \cdot s + b_2} \tag{3.9}$$

We can generalize these results for the first-order factor below.

- The gain constant is unchanged.
- The s-term coefficients become
  $a_1 = A_1$, $b_1 = B_1$
- The constant term coefficients become
  $a_2 = A_2\,\omega_o$, $b_2 = B_2\,\omega_o$

3.1.2 Handling a Second-Order Factor

In the case of the quadratic terms that are used to describe our coefficients, the unnormalization process is shown in Equation 3.10 and 3.11.

$$H(s) = \left.\frac{A_0 \cdot S^2 + A_1 \cdot S + A_2}{B_0 \cdot S^2 + B_1 \cdot S + B_2}\right|_{S=s/\omega_o} = \frac{A_0 \cdot (s/\omega_o)^2 + A_1 \cdot (s/\omega_o) + A_2}{B_0 \cdot (s/\omega_o)^2 + B_1 \cdot (s/\omega_o) + B_2} \quad (3.10)$$

$$H(s) = \frac{A_0 \cdot s^2 + A_1 \cdot \omega_o \cdot s + A_2 \cdot \omega_o^2}{B_0 \cdot s^2 + B_1 \cdot \omega_o \cdot s + B_2 \cdot \omega_o^2} = \frac{a_0 \cdot s^2 + a_1 \cdot s + a_2}{b_0 \cdot s^2 + b_1 \cdot s + b_2} \quad (3.11)$$

The coefficient A_0 will be 1 or 0. A value of 1 will be present only if an inverse Chebyshev or elliptic approximation is being unnormalized, while a 0 will be used for Chebyshev and Butterworth. A_1 will normally be 0 for all approximations, but is included for completeness of the derivation in the event we would like to use any of our work at a later time when complex conjugate zeros will occur off the $j\omega$ axis. B_0 will typically be 1 for all cases, but is retained for generality. By observation, we can determine the following relationships which can be used in our C code.

- The gain constant is unchanged.
- The s^2-term coefficients become
  $a_0 = A_0$, $b_0 = B_0$
- The s-term coefficients become
  $a_1 = A_1\,\omega_o$, $b_1 = B_1\,\omega_o$
- The constant term coefficients become
  $a_2 = A_2\omega_o^2$, $b_2 = B_2\omega_o^2$

Complete numerical examples of the lowpass unnormalization process are now in order.

Example 3.1 Unnormalized Inv. Chebyshev Lowpass Filter

Problem: Determine the transfer function for an inverse Chebyshev lowpass filter to satisfy the specifications:

$a_{\text{pass}} = -0.25$ dB, $a_{\text{stop}} = -38.0$ dB,

$\omega_{\text{pass}} = 600$ rad/sec, and $\omega_{\text{stop}} = 1000$ rad/sec

Solution: Using the material of Section 2.4, the important values for this example and the normalized transfer function are listed below. The unnormalized transfer function is then determined by making the substitution $S = s / \omega_0$ and using the relationships just developed.

$\Omega_r = 1.667$ $\quad n = 5.90$ (6th order) $\quad \omega_0 = 1000.0$ rad/sec

$$H_{I,6}(S) = \frac{(S^2 + 1.0718)\cdot(S^2 + 2.0000)\cdot(S^2 + 14.9282)}{(S^2 + 0.2679\cdot S + 0.5455)\cdot(S^2 + 0.9583\cdot S + 0.7142)\cdot(S^2 + 1.8952\cdot S + 1.0340)}$$

$$H_{I,6}(s) = \frac{(s^2 + 1071.80\cdot 10^3)\cdot(s^2 + 2000.0\cdot 10^3)\cdot(s^2 + 14928.2\cdot 10^3)}{(s^2 + 267.91\cdot s + 545.5\cdot 10^3)\cdot(s^2 + 958.3\cdot s + 714.2\cdot 10^3)\cdot(s^2 + 1895.2\cdot s + 1034.0\cdot 10^3)}$$

Example 3.2 Unnormalized Butterworth Lowpass Filter

Problem: Determine the transfer function for a Butterworth lowpass filter to satisfy the following specifications:

$a_{\text{pass}} = -0.5$ dB, $a_{\text{stop}} = -21$ dB, $f_{\text{pass}} = 1000$ Hz, and $f_{\text{stop}} = 2000$ Hz

Solution: Using the material of Section 2.2, the important values for this example and the normalized transfer function are listed below.

$\Omega_r = 2.0$ $\quad n = 5.00$ (5th order) $\quad \omega_0 = 6283.19$ rad/sec

$$H_{B,5}(S) = \frac{1.23412\cdot 1.52305\cdot 1.52305}{(S + 1.23412)\cdot(S^2 + 1.99685\cdot S + 1.52305)\cdot(S^2 + 0.76273\cdot S + 1.52305)}$$

The unnormalized transfer function is then determined by making the substitution $S = s / \omega_0$ and using the relationships just developed.

$$H_{B,5}(s) = \frac{7754.21\cdot 6.01277\cdot 10^7\cdot 6.01277\cdot 10^7}{(s + 7754.21)\cdot(s^2 + 12546.6s + 6.01277\cdot 10^7)\cdot(s^2 + 4792.36 + 6.01227\cdot 10^7)}$$

We can also use FILTER to design either of the lowpass filters just described, but in this case we'll pick the Butterworth filter. The coefficients and response of this filter are shown in Figure 3.2 and Figure 3.3. FILTER also calculates the passband and stopband edge frequency gains (although not shown) as –0.5 dB and –21.0 dB, which meet the specifications.

```
                         Butterworth Lowpass Filter
                  Filter implementation is:      Analog
                  Filter approximation is:       Butterworth
                  Filter selectivity is:         Lowpass
                  Passband gain         (dB):    -0.50
                  Stopband gain         (dB):    -21.00
                  Passband frequency    (Hz):    1000.00
                  Stopband frequency    (Hz):    2000.00

                                 Order = 5
                         Overall Gain = 1.00000000e+000
      Numerator Coefficients                    Denominator Coefficients
  [S^2 +       S          +       ()      ]  [S^2 +       S          +      ()      ]
  =================================  =================================
1 0.0 0.00000000e+00 7.75420568e+003  0.0 1.00000000e+00 7.75420568e+003
2 0.0 0.00000000e+00 6.01277057e+007  1.0 4.79236267e+03 6.01277057e+007
3 0.0 0.00000000e+00 6.01277057e+007  1.0 1.25465683e+04 6.01277057e+007
```

Figure 3.2 Filter coefficients for Example 3.2.

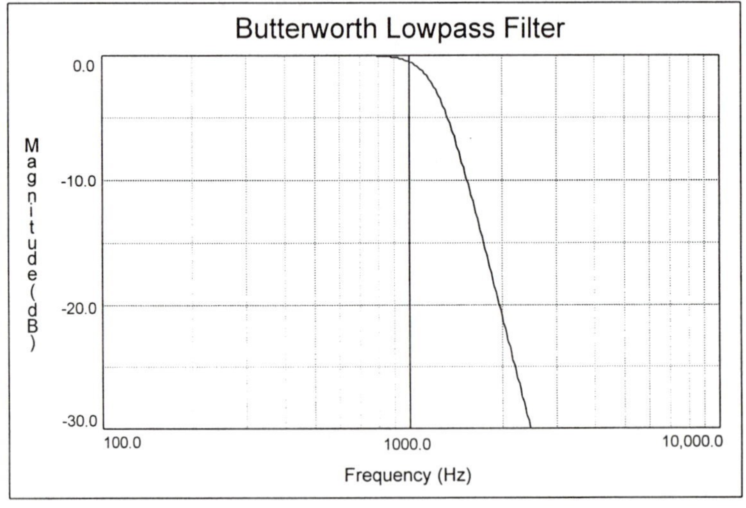

Figure 3.3 Filter magnitude response for Example 3.2.

3.2 UNNORMALIZED HIGHPASS APPROXIMATION FUNCTIONS

The normalized lowpass approximation can also be used to generate the approximation function for a highpass filter. The calculation for the ratio frequency Ω_r is based on the ratio of passband to stopband frequencies as shown in Equation 3.12. This is the reciprocal of the lowpass case, but since $\omega_{pass} > \omega_{stop}$, the result still produces a value greater than 1. The Ω_r ratio always produces a value greater than 1 (as we will see in later sections as well) and as the value gets larger and larger, the order of the filter will reduce as long as other characteristics remain the same.

$$\Omega_{rH} = \frac{\omega_{pass}}{\omega_{stop}} = \frac{f_{pass}}{f_{stop}} \tag{3.12}$$

Once the normalized lowpass approximation is determined based on the order, we can unnormalize the lowpass transfer function using an appropriate unnormalization substitution. In the case of the highpass filter, the unnormalization substitution is given in Equation 3.13. As in the lowpass case, ω_o will take on the value of ω_{pass} except for the inverse Chebyshev approximation where it will have the value of ω_{stop}.

$$S_H = \frac{\omega_o}{s} \tag{3.13}$$

3.2.1 Handling a First-Order Factor

For the first-order case, we start with Equation 3.14 and make the substitution of Equation 3.13. The final result is then shown in Equation 3.15. In this unnormalization case, we see that there is a gain adjustment (A_2 / B_2) which must be considered.

$$H(s) = \left.\frac{A_1 \cdot S + A_2}{B_1 \cdot S + B_2}\right|_{S=\omega_o/s} = \frac{A_1 \cdot (\omega_o/s) + A_2}{B_1 \cdot (\omega_o/s) + B_2} \tag{3.14}$$

$$H(s) = \frac{A_2}{B_2} \cdot \frac{s + (A_1 / A_2) \cdot \omega_o}{s + (B_1 / B_2) \cdot \omega_o} = \frac{A_2}{B_2} \cdot \frac{a_1 \cdot s + a_2}{b_1 \cdot s + b_2} \tag{3.15}$$

From careful observation we can draw the following information from these equations.

- The gain constant is multiplied by A_2 / B_2.
- The s-term coefficients become
  $a_1 = 1$, $b_1 = 1$
- The constant term coefficients become
  $a_2 = (A_1 / A_2)\,\omega_o$, $b_2 = (B_1 / B_2)\,\omega_o$

3.2.2 Handling a Second-Order Factor

In the case of this highpass unnormalization process, the second-order factors will be unnormalized in the manner shown in Equation 3.16,

$$H(s) = \left.\frac{A_0 \cdot S^2 + A_1 \cdot S + A_2}{B_0 \cdot S^2 + B_1 \cdot S + B_2}\right|_{S=\omega_o/s} = \frac{A_0 \cdot (\omega_o/s)^2 + A_1 \cdot (\omega_o/s) + A_2}{B_0 \cdot (\omega_o/s)^2 + B_1 \cdot (\omega_o/s) + B_2} \qquad (3.16)$$

which can be simplified to produce

$$H(s) = \frac{A_2}{B_2} \cdot \frac{s^2 + (A_1/A_2)\cdot\omega_o \cdot s + (A_0/A_2)\cdot\omega_o^2}{s^2 + (B_1/B_2)\cdot\omega_o \cdot s + (B_0/B_2)\cdot\omega_o^2} = \frac{A_2}{B_2} \cdot \frac{a_0 \cdot s^2 + a_1 \cdot s + a_2}{b_0 \cdot s^2 + b_1 \cdot s + b_2} \qquad (3.17)$$

Notice that if A_0 and A_1 are both zero (which will be the case for Butterworth and Chebyshev approximations), then a_1 and a_2 will be zero, leaving only an s^2-term in the numerator. A_2 will never be zero in a normalized approximation function. We can summarize the results of the unnormalization below.

- The gain constant is multiplied by A_2 / B_2.
- The s^2-term coefficients become
  $a_0 = 1$, $b_0 = 1$
- The s-term coefficients become
  $a_1 = (A_1 / A_2)\,\omega_o$, $b_1 = (B_1 / B_2)\,\omega_o$
- The constant term coefficients become
  $a_2 = (A_0 / A_2)\omega_o^2$, $b_2 = (B_0 / B_2)\omega_o^2$

Numerical examples of the highpass unnormalization process can now be used to better illustrate the process.

Example 3.3 Unnormalized Elliptic Highpass Filter

Problem: Determine the transfer function for an elliptic highpass filter to satisfy the following specifications: $a_{pass} = -0.5$ dB, $a_{stop} = -45.0$ dB,

$\omega_{pass} = 3000$ rad/sec, and $\omega_{stop} = 2000$ rad/sec

Solution: Using the material of Section 2.5, the important values for this example and the normalized transfer function are listed below.

$\Omega_r = 1.5$ $n = 4.61$ (5th order) $\omega_o = 3000.0$ rad/sec

$$H_{E,5}(S) = \frac{0.045066 \cdot 0.42597 \cdot (S^2 + 5.43764) \cdot (S^2 + 2.42551)}{(S + 0.42597) \cdot (S^2 + 0.57023 \cdot S + 0.57601) \cdot (S^2 + 0.16346 \cdot S + 1.03189)}$$

The unnormalized transfer function is then determined by making the substitution $S = \omega_o / s$ and using the relationships just developed.

$$H_{E,5}(s) = \frac{s \cdot (s^2 + 1.65513 \cdot 10^6) \cdot (s^2 + 3.71055 \cdot 10^6)}{(s + 7042.75) \cdot (s^2 + 2969.91s + 1.56247 \cdot 10^7) \cdot (s^2 + 475.228 + 8.72183 \cdot 10^6)}$$

Example 3.4 Unnormalized Chebyshev Highpass Filter

Problem: Determine the transfer function for a Chebyshev highpass filter to satisfy the following specifications:

$a_{pass} = -1.5$ dB, $a_{stop} = -40$ dB, $f_{pass} = 2000$ Hz, and $f_{stop} = 800$ Hz

Solution: Using the material of Section 2.3, the important values for this example and the normalized transfer function are listed below.

$\Omega_r = 2.5$ $n = 3.66$ (4th order) $\omega_o = 12{,}566.4$ rad/sec

$$H_{C,4}(S) = \frac{0.84140 \cdot 0.95046 \cdot 0.24336}{(S^2 + 0.23826 \cdot S + 0.95046) \cdot (S^2 + 0.57521 \cdot S + 0.24336)}$$

The unnormalized transfer function is then determined by making the substitution $S = \omega_o / s$ and using the relationships just developed.

$$H_{C,4}(s) = \frac{0.84140 \cdot s^2 \cdot s^2}{(s^2 + 3150.13 \cdot s + 1.66144 \cdot 10^8) \cdot (s^2 + 2970.27 \cdot s + 6.48899 \cdot 10^8)}$$

The results of using FILTER to design the filter of Example 3.4 are illustrated in Figures 3.4 and 3.5, which show the coefficients and magnitude response. FILTER also calculates (not shown) the passband edge frequency gain to be −1.5 dB and the stopband edge frequency gain to be −44.6 dB, which meets or exceeds the specification.

```
                         Chebyshev Highpass Filter

                Filter implementation is:      Analog
                Filter approximation is:       Chebyshev
                Filter selectivity is:         Highpass
                Passband gain        (dB):     -1.50
                Stopband gain        (dB):     -40.00
                Passband frequency   (Hz):     2000.00
                Stopband frequency   (Hz):     800.00

                              Order = 4
                       Overall Gain = 8.41395142e-001
        Numerator Coefficients                 Denominator Coefficients
  [S^2 +       S          +       ()       ]  [S^2 +       S          +       ()       ]
  =================================  =================================
1 1.0 0.00000000e+00 0.00000000e+000  1.0 3.15012808e+03 1.66143895e+008
2 1.0 0.00000000e+00 0.00000000e+000  1.0 2.97027255e+04 6.48898536e+008
```

Figure 3.4 Filter coefficients for Example 3.4.

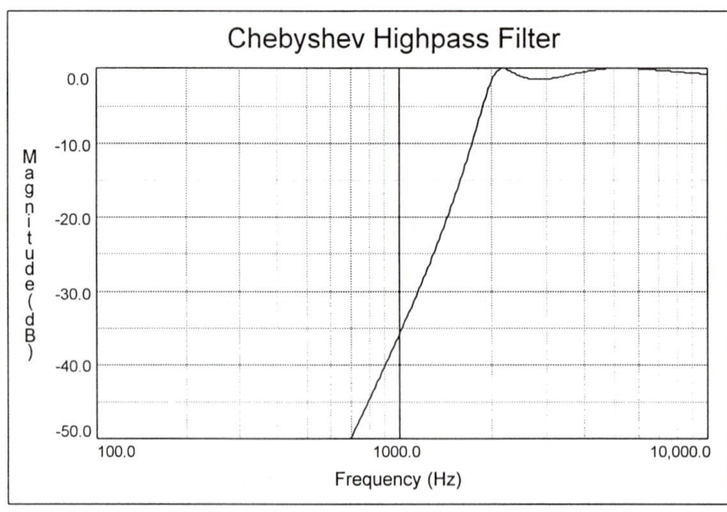

Figure 3.5 Filter magnitude response for Example 3.4.

3.3 UNNORMALIZED BANDPASS APPROXIMATION FUNCTIONS

In the case of a bandpass unnormalization, Ω_r will be defined in Equation 3.18. Note that for this case, Ω_r will be greater than 1 as has been the case for the lowpass and highpass unnormalization.

$$\Omega_{rP} = \frac{\omega_{stop2} - \omega_{stop1}}{\omega_{pass2} - \omega_{pass1}} = \frac{f_{stop2} - f_{stop1}}{f_{pass2} - f_{pass1}} \tag{3.18}$$

After determining the normalized lowpass approximation using the order we determined from the bandpass specifications, we can unnormalize the lowpass function into a bandpass function. To accomplish this we use the substitution given in Equation 3.19.

$$S_P = \frac{s^2 + \omega_o^2}{BW \cdot s} \tag{3.19}$$

where for all approximations except the inverse Chebyshev,

$$\omega_o = \sqrt{\omega_{pass1} \cdot \omega_{pass2}} \tag{3.20}$$

$$BW = \omega_{pass2} - \omega_{pass1} \tag{3.21}$$

For the inverse Chebyshev case these values are defined as (another example of the opposite nature of this approximation)

$$\omega_o = \sqrt{\omega_{stop1} \cdot \omega_{stop2}} \tag{3.22}$$

$$BW = \omega_{stop2} - \omega_{stop1} \tag{3.23}$$

In order to provide an accurate value of Ω_r in Equation 3.18, the stopband and passband edge frequencies must be symmetrically spaced on either side of ω_o. The simplest way to test for this is to check to see that the relationship of Equation 3.24 is satisfied. If this equation is not satisfied, the larger side must be reduced to form an equality by increasing ω_{stop1} or decreasing ω_{stop2}. This will tighten the restrictions, therefore the original specifications will still be met, and an accurate order can be calculated. (If there is an extreme inequality, other measures can be used to implement the filter. For example, an additional lowpass or highpass filter can be added to provide the required selectivity.)

$$\frac{\omega_{\text{pass1}}}{\omega_{\text{stop1}}} = \frac{\omega_{\text{stop2}}}{\omega_{\text{pass2}}} \tag{3.24}$$

As indicated by Equation 3.19, the unnormalization process will result in a bandpass approximation function which has twice the order of the lowpass function used to generate it. This seems reasonable when we consider that a bandpass filter must provide a transition from a stopband to a passband (like a highpass filter) and another transition from a passband to a stopband (like a lowpass filter). The resulting function must therefore be twice the order of the original lowpass function on which it is based.

3.3.1 Handling a First-Order Factor

For a first-order factor in the lowpass approximation function, Equation 3.25 shows how the substitution of Equation 3.19 is made.

$$H(s) = \left.\frac{A_1 \cdot S + A_2}{B_1 \cdot S + B_2}\right|_{S=(s^2+\omega_o^2)/(BW\cdot s)} = \frac{A_1 \cdot [(s^2+\omega_o^2)/(BW\cdot s)] + A_2}{B_1 \cdot [(s^2+\omega_o^2)/(BW\cdot s)] + B_2} \tag{3.25}$$

And after some simplification we have the result in Equation 3.26. The relationships between the coefficients are shown. Note that if $A_1 = 0$, as will normally be the case, the numerator will only have an s-term present.

$$H(s) = \frac{A_1 \cdot s^2 + A_2 \cdot BW \cdot s + A_1 \cdot \omega_o^2}{B_1 \cdot s^2 + B_2 \cdot BW \cdot s + B_1 \cdot \omega_o^2} = \frac{a_0 \cdot s^2 + a_1 \cdot s + a_2}{b_0 \cdot s^2 + b_1 \cdot s + b_2} \tag{3.26}$$

- The gain constant is unchanged.
- The s^2-term bandpass coefficients become
  $a_0 = A_1,\ b_0 = B_1$
- The s-term bandpass coefficients become
  $a_1 = A_2\,BW,\ b_1 = B_2\,BW$
- The constant term bandpass coefficients become
  $a_2 = A_1\omega_o^2,\ b_2 = B_1\omega_o^2$

3.3.2 Handling a Second-Order Factor

Unnormalizing a second-order factor is a bit more of a challenge. When the substitution variable S_P of Equation 3.19 is inserted into a second-order lowpass approximation, a fourth-order factor results. What do we do with a fourth-order factor? All of our development to this point is based on quadratic factors and with good reason. They represent a complex conjugate pair and they will be used to efficiently implement the filters in later chapters. We could factor the fourth-order, but this would require a numerical algorithm which is time-consuming and not always accurate. There is another directed procedure which can be used.

If we factor the lowpass approximation quadratic into two complex conjugate factors before making the substitution of Equation 3.19, the result after the substitution and simplification is two quadratic equations. However, each of these quadratics would have a complex coefficient which would mean that they could not be implemented directly. But if these two quadratics are again factored, we will find two sets of complex conjugate pairs within the set of four factors. These complex conjugate pairs could then be combined to produce two quadratics which have all real coefficients.

Perhaps an easy example is in order. Consider the transfer function shown in Equation 3.27 which has already been factored.

$$H(S) = \frac{2}{S^2 + 2 \cdot S + 2} = \frac{2}{(S + 1 + j1) \cdot (S + 1 - j1)} \qquad (3.27)$$

Now if we assume that $\omega_o = 1$ and $BW = 1$, we can substitute $S = (s^2 + 1) / s$ and simplify to produce the following.

$$H(s) = \frac{2 \cdot s^2}{[s^2 + (1 + j1) \cdot s + 1] \cdot [s^2 + (1 - j1) \cdot s + 1]} \qquad (3.28)$$

The roots of the first quadratic can be determined to be

$$s_{1,2} = \frac{-(1 + j1) \pm (0.486 + j2.058)}{2} = (-0.257 + j0.529),\ (-0.743 - j1.53) \qquad (3.29a)$$

and the two roots of the second quadratic pair up with the first.

$$s_{3,4} = (-0.257 - j0.529),\ (-0.743 + j1.53) \qquad (3.29b)$$

The resulting transfer function can then be written as Equation 3.30 by combining the complex conjugate roots from each quadratic.

$$H(s) = \frac{2 \cdot s^2}{(s^2 + 0.514 \cdot s + 0.346) \cdot (s^2 + 1.486 \cdot s + 2.890)} \tag{3.30}$$

This algorithm for finding the two quadratics in the bandpass approximation from the single quadratic in the lowpass function will be used as the standard method in this section. Unfortunately, it is very difficult to define the final bandpass coefficients in terms of only the initial lowpass coefficients because of the complexity of expressions. However, if we use a few intermediate variables, the process should be able to be demonstrated without too much confusion.

We start with the general expression for the normalized lowpass second-order factor shown in Equation 3.31 in normal and factored form. The A_0 and B_0 coefficients have been omitted for clarity since they will be assumed to be 1 in this case. The p's and z's represent the complex poles and zeros, respectively. (An asterisk indicates the complex conjugate value.)

$$H(s) = \left.\frac{S^2 + A_1 \cdot S + A_2}{S^2 + B_1 \cdot S + B_2}\right|_{S=(s^2+\omega_o^2)/(BW \cdot s)} = \left.\frac{(S+z_1)\cdot(S+z_1^*)}{(S+p_1)\cdot(S+p_1^*)}\right|_{S=(s^2+\omega_o^2)/(BW \cdot s)} \tag{3.31}$$

After making the indicated substitution and simplifying, we have

$$H(s) = \frac{(s^2 + BW \cdot z_1 \cdot s + \omega_o^2)\cdot(s^2 + BW \cdot z_1^* \cdot s + \omega_o^2)}{(s^2 + BW \cdot p_1 \cdot s + \omega_o^2)\cdot(s^2 + BW \cdot p_1^* \cdot s + \omega_o^2)} \tag{3.32}$$

Each one of the quadratic factors in Equation 3.32 can now be factored into first-order factors as indicated in Equation 3.33. Note that the constants from the z_1 quadratic are labeled with an a and b, while the complex conjugates use c and d. The denominator uses the same designations.

$$H(s) = \frac{(s + z_{1a})\cdot(s + z_{1b})\cdot(s + z_{1c})\cdot(s + z_{1d})}{(s + p_{1a})\cdot(s + p_{1b})\cdot(s + p_{1c})\cdot(s + p_{1d})} \tag{3.33}$$

Now, by matching the complex conjugate pairs, we can reconstruct two quadratics with real coefficients in both the numerator and denominator.

$$H(s) = \frac{(s^2 + a_1 \cdot s + a_2) \cdot (s^2 + a_4 \cdot s + a_5)}{(s^2 + b_1 \cdot s + b_2) \cdot (s^2 + b_4 \cdot s + b_5)} \tag{3.34}$$

The results shown in Equation 3.34 are valid for the inverse Chebyshev and elliptic approximation functions which use zeros on the $j\omega$ axis. However, for the Butterworth and Chebyshev approximation functions, the result will be somewhat different. Equation 3.35 shows the starting point for this development. After substitution and simplification Equation 3.36 results. Then following the same basic steps as in the previous derivation, Equation 3.37 eventually emerges.

$$H(S) = \left.\frac{A_2}{S^2 + B_1 \cdot S + B_2}\right|_{S=(s^2+\omega_o^2)/(BW \cdot s)} = \left.\frac{A_2}{(S + p_1) \cdot (S + p_1^*)}\right|_{S=(s^2+\omega_o^2)/(BW \cdot s)} \tag{3.35}$$

$$H(s) = \frac{A_2 \cdot BW^2 \cdot s^2}{(s^2 + BW \cdot p_1 \cdot s + \omega_o^2) \cdot (s^2 + BW \cdot p_1^* \cdot s + \omega_o^2)} \tag{3.36}$$

$$H(s) = \frac{A_2 \cdot BW^2 \cdot s^2}{(s^2 + b_1 \cdot s + b_2) \cdot (s^2 + b_4 \cdot s + b_5)} \tag{3.37}$$

Now a couple of examples illustrating the bandpass design process are in order.

Example 3.5 Unnormalized Butterworth Bandpass Filter

Problem: Determine the transfer function for a Butterworth bandpass filter to satisfy the following specifications:

$a_{pass} = -1.0$ dB, $a_{stop} = -21$ dB, $f_{pass1} = 300$ Hz, $f_{pass2} = 3000$ Hz, $f_{stop1} = 50$ Hz, and $f_{stop2} = 9000$ Hz

Solution: Using the material of Section 2.2, the important values for this example and the normalized transfer function are listed below. In this case, f_{stop1} must be changed to 100 Hz to provide symmetry. The function is shown with quadratics in factored form as indicated by the (2) superscript.

$\Omega_r = 3.3$ $\quad$ $n = 2.59$ (3rd order)
$\omega_0 = 5960.8$ rad/sec $\quad$ $BW = 16{,}965$ rad/sec

$$H_{B,3}(S) = \frac{1.25258 \cdot 1.56895}{(S + 1.25258) \cdot (S + 0.62629 \pm j1.08476)^{(2)}}$$

After making the substitution of Equation 3.19 and factoring again, the following equation emerges.

$$H_{B,6}(s) = \frac{1.25258 \cdot 1.56895 \cdot (16965 \cdot s)^3}{(s^2 + 21249s + 3.5531 \cdot 10^7) \cdot (s + 716.08 \pm j1433.5)^{(2)} (s + 9908.7s \pm j19836)^{(2)}}$$

After simplification, the following transfer function results.

$$H_{B,6}(s) = \frac{9.59570 \cdot 10^{12} \cdot s^3}{(s^2 + 21249s + 3.5531 \cdot 10^7) \cdot (s^2 + 1432.2s + 2.5677 \cdot 10^6) \cdot (s^2 + 19817 + 4.9165 \cdot 10^8)}$$

Example 3.6 Unnormalized Inv. Chebyshev Bandpass Filter

Problem: Determine the transfer function for an inverse Chebyshev bandpass filter to satisfy the following specifications:

$a_{pass} = -0.5$ dB, $a_{stop} = -33$ dB, $\omega_{pass1} = 100$ rad/sec,
$\omega_{pass2} = 200$ rad/sec, $\omega_{stop1} = 50$ rad/sec, and $\omega_{stop2} = 400$ rad/sec

Solution: Using the material of Section 2.4, the important values for this example and the normalized transfer function are listed below. The transfer function is shown with quadratics in factored form as indicated by the (2) superscript.

$\Omega_r = 3.5$ $\quad$ $n = 2.88$ (3rd order)
$\omega_0 = 141.421$ rad/sec $\quad$ $BW = 350.00$ rad/sec

$$H_{I,3}(S) = \frac{0.47098 \cdot (S \pm j1.15470)^{(2)}}{(S + 0.47098) \cdot (S + 0.20190 \pm j0.38655)^{(2)}}$$

After making the substitution of Equation 3.19 and factoring again, the following equation emerges.

$$H_{I,6}(s)=\frac{(164.84\cdot s)\cdot(s\pm j448.72)^{(2)}\cdot(s\pm j44.572)^{(2)}}{(s^2+164.84\cdot s+20{,}000)\cdot(s+50.901\pm j221.17)^{(2)}(s+19.765\pm j85.880)^{(2)}}$$

After simplification, the following transfer function results.

$$H_{I,6}(s)=\frac{(164.84\cdot s)\cdot(s^2+201350)\cdot(s^2+1986.6)}{(s^2+164.84\cdot s+20{,}000)\cdot(s^2+101.80\cdot s+51{,}507)\cdot(s^2+39.529\cdot s+7765.9)}$$

We can use FILTER to design the inverse Chebyshev bandpass filter of Example 3.6. The coefficients for this design are shown in Figure 3.6 with the magnitude response shown in Figure 3.7. Notice that a couple of the coefficients are very small (10^{-14}) and can interpreted as zero. FILTER also determines (not shown) that the stopband edge frequency gain is –33.0 dB while the passband edge frequency gain is –0.32 dB. These gains meet or exceed the specifications.

```
                    Inv. Chebyshev Bandpass Filter

                Filter implementation is:        Analog
                Filter approximation is:         Inv Cheby
                Filter selectivity is:           Bandpass
                Passband gain             (dB):  -0.50
                Stopband gain - lower (dB):      -33.00
                Stopband gain - upper (dB):      -33.00
                Passband freq - lower (Hz):      15.92
                Passband freq - upper (Hz):      31.83
                Stopband freq - lower (Hz):      7.96
                Stopband freq - upper (Hz):      63.66

                            Order = 6
                   Overall Gain = 1.42635925e-001

       Numerator Coefficients                 Denominator Coefficients
  [S^2 +      S          +       ()       ]  [S^2 +      S          +       ()      ]
  =================================  =================================
1 0.0 1.64842513e+02 0.00000000e+000   1.0 1.64842513e+02 1.99999999e+004
2 1.0 -5.4950136e-14 2.01346705e+005   1.0 3.95288944e+01 7.76590294e+003
3 1.0 5.45826023e-15 1.98662300e+003   1.0 1.01801154e+02 5.15072101e+004
```

Figure 3.6 Filter coefficients for Example 3.6.

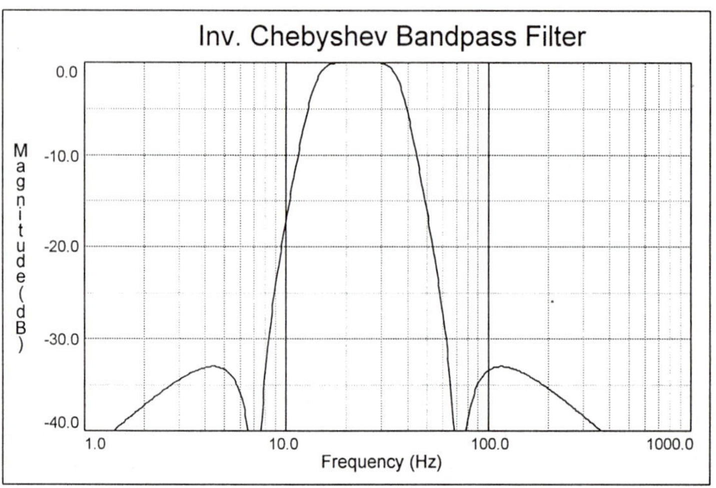

Figure 3.7 Filter magnitude response for Example 3.6.

3.4 Unnormalized Bandstop Approximation Functions

We will find that the unnormalization of the lowpass normalized function into a bandstop approximation is very similar to the bandpass case. The first step in the procedure is to determine the order required from the lowpass approximation based on the bandstop specifications. The value of Ω_r to use in the bandstop case is shown in Equation 3.38, which is the reciprocal of the bandpass case. We notice again that Ω_r will be greater than one.

$$\Omega_r = \frac{\omega_{\text{pass2}} - \omega_{\text{pass1}}}{\omega_{\text{stop2}} - \omega_{\text{stop1}}} = \frac{f_{\text{pass2}} - f_{\text{pass1}}}{f_{\text{stop2}} - f_{\text{stop1}}} \tag{3.38}$$

As in the bandpass case, Equation 3.24 must be satisfied in order to get an accurate value of Ω_r, except in this case either ω_{pass1} must be increased or ω_{pass2} must be decreased to achieve equality. After finding the order, we can unnormalize the lowpass function into a bandstop function using the substitution given in Equation 3.39.

$$S_S = \frac{BW \cdot s}{s^2 + \omega_o^2} \tag{3.39}$$

As before, all approximations except the inverse Chebyshev will define

$$\omega_o = \sqrt{\omega_{pass1} \cdot \omega_{pass2}} \tag{3.40}$$

$$BW = \omega_{pass2} - \omega_{pass1} \tag{3.41}$$

while for the inverse Chebyshev case

$$\omega_o = \sqrt{\omega_{stop1} \cdot \omega_{stop2}} \tag{3.42}$$

$$BW = \omega_{stop2} - \omega_{stop1} \tag{3.43}$$

The resultant bandstop approximation function will be twice the order of the normalized lowpass function just as in the bandpass case. The bandstop filter is in effect implementing both a lowpass and highpass filter and therefore requires twice the order.

3.4.1 Handling a First-Order Factor

Equation 3.44 shows how a first-order factor is unnormalized into a second-order factor using the substitution of Equation 3.39.

$$H(S) = \left.\frac{A_1 \cdot S + A_2}{B_1 \cdot S + B_2}\right|_{S=(BW\cdot s)/(s^2+\omega_o^2)} = \frac{A_1 \cdot [(BW\cdot s)/(s^2+\omega_o^2)] + A_2}{B_1 \cdot [(BW\cdot s)/(s^2+\omega_o^2)] + B_2} \tag{3.44}$$

Equation 3.45 shows the result after simplification followed by the observations which can be made for this case.

$$H(s) = \frac{A_2}{B_2} \cdot \frac{s^2 + (A_1 / A_2) \cdot BW \cdot s + \omega_o^2}{s^2 + (B_1 / B_2) \cdot BW \cdot s + \omega_o^2} = \frac{a_0 \cdot s^2 + a_1 \cdot s + a_2}{b_0 \cdot s^2 + b_1 \cdot s + b_2} \tag{3.45}$$

- The gain constant is multiplied by A_2 / B_2.
- The s^2-term bandstop coefficients become
  $a_0 = 1,\ b_0 = 1$
- The s-term bandstop coefficients become
  $a_1 = (A_1 / A_2)\ BW,\ b_1 = (B_1 / B_2)\ BW$
- Constant term bandstop coefficients become
  $a_2 = \omega_o^2,\ b_2 = \omega_o^2$

3.4.2 Handling a Second-Order Factor

In order to unnormalize a second-order factor we experience the same problems as in the bandpass case. A direct substitution would give us a fourth-order transfer function, which is not what we want. However, the methodology used in the bandpass case does work in this case as well. The procedure will be outlined below for the bandstop case which has a few differences due to a different substitution factor.

Starting at the same point as with the bandpass case, Equation 3.46 shows the result of the factoring of the initial quadratics.

$$H(s) = \left.\frac{S^2 + A_1 \cdot S + A_2}{S^2 + B_1 \cdot S + B_2}\right|_{S=(BW \cdot s)/(s^2+\omega_o^2)} = \left.\frac{(S+z_1)\cdot(S+z_1^*)}{(S+p_1)\cdot(S+p_1^*)}\right|_{S=(BW \cdot s)/(s^2+\omega_o^2)} \tag{3.46}$$

Equation 3.47 results after the substitution and simplification.

$$H(s) = \frac{z_1 \cdot z_1^*}{p_1 \cdot p_1^*} \cdot \frac{[s^2 + (BW/z_1)\cdot s + \omega_o^2]\cdot[s^2 + (BW/z_1^*)\cdot s + \omega_o^2]}{[s^2 + (BW/p_1)\cdot s + \omega_o^2]\cdot[s^2 + (BW/p_1^*)\cdot s + \omega_o^2]} \tag{3.47}$$

The initial gain factor of Equation 3.47 can be shown to be $A_2 \,/\, B_2$. The quadratic factors can be factored into first-order factors as indicated in Equation 3.48.

$$H(s) = \frac{A_2}{B_2} \cdot \frac{(s+z_{1a})\cdot(s+z_{1b})\cdot(s+z_{1c})\cdot(s+z_{1d})}{(s+p_{1a})\cdot(s+p_{1b})\cdot(s+p_{1c})\cdot(s+p_{1d})} \tag{3.48}$$

We then reconstruct two quadratics in the numerator and denominator by matching the complex conjugate pairs.

$$H(s) = \frac{A_2}{B_2} \cdot \frac{(s^2 + a_1 \cdot s + a_2)\cdot(s^2 + a_4 \cdot s + a_5)}{(s^2 + b_1 \cdot s + b_2)\cdot(s^2 + b_4 \cdot s + b_5)} \tag{3.49}$$

This result is valid for the rational approximation functions (inverse Chebyshev and elliptic), but for the all-pole approximations (Butterworth and Chebyshev), we must develop a slightly different version. Equations 3.50 and 3.51 show the factoring and substitution of Equation 3.39. After the quadratics of Equation 3.51 are factored and the matching complex conjugate terms are combined, the final form of Equation 3.52 results.

$$H(S) = \left.\frac{A_2}{S^2 + B_1 \cdot S + B_2}\right|_{S=(BW \cdot s)/(s^2+\omega^2)} = \left.\frac{A_2}{(S+p_1)\cdot(S+p_1^*)}\right|_{S=(BW \cdot s)/(s^2+\omega^2)} \tag{3.50}$$

$$H(s) = \frac{A_2}{p_1 \cdot p_1^*} \cdot \frac{(s^2+\omega_o^2)^2}{[s^2+(BW/p_1)\cdot s+\omega_o^2]\cdot[s^2+(BW/p_1^*)\cdot s+\omega_o^2]} \tag{3.51}$$

$$H(s) = \frac{A_2}{B_2} \cdot \frac{(s^2+\omega_o^2)^2}{(s^2+b_1 \cdot s+b_2)\cdot(s^2+b_4 \cdot s+b_5)} \tag{3.52}$$

Complete numerical examples of the bandstop unnormalization process are given next.

Example 3.7 Unnormalized Chebyshev Bandstop Filter

Problem: Determine the transfer function for a Chebyshev bandstop filter to satisfy the following specifications:

$\omega_{pass1} = 3000$ rad/sec, $\omega_{pass2} = 24{,}000$ rad/sec, $\omega_{stop1} = 6000$ rad/sec,

$\omega_{stop2} = 12{,}000$ rad/sec, $a_{pass} = -1.0$ dB, and $a_{stop} = -35$ dB

Solution: Using the material of Section 2.3, the important values for this example and the normalized transfer function are listed below. The transfer function is shown with quadratics in factored form as shown by the (2) superscript.

$\Omega_r = 3.5$ $\qquad$ $n = 2.80$ (3rd order)

$\omega_o = 8485.3$ rad/sec $\qquad$ $BW = 21{,}000$ rad/sec

$$H_{C,3}(S) = \frac{0.49417 \cdot 0.99421}{(S+0.49417)\cdot(S+0.24709 \pm j0.96600)^{(2)}}$$

After making the substitution of Equation 3.39 and factoring again, the following equation emerges, which can be simplified into the final result.

$$H_{C,6}(s) = \frac{(s^2+7.200\cdot 10^7)^3}{(s^2+42{,}495\cdot s+7.2000\cdot 10^7)\cdot(s+587.55\pm j2964.6)^{(2)}(s+4631.5\pm j23{,}369)^{(2)}}$$

$$H_{C,6}(s) = \frac{(s^2 + 7.200 \cdot 10^7)^3}{(s^2 + 42{,}495 \cdot s + 7.2000 \cdot 10^7) \cdot (s^2 + 1175.1 \cdot s + 9.1340 \cdot 10^6)(s^2 + 9263.0 \cdot s + 5.6755 \cdot 10^8)}$$

Example 3.8 Unnormalized Elliptic Bandstop Filter

Problem: Determine the transfer function for an elliptic bandstop filter to satisfy the following specifications:

$a_{\text{pass}} = -0.3$ dB, $a_{\text{stop}} = -50$ dB, $f_{\text{pass1}} = 50$ Hz, $f_{\text{pass2}} = 72$ Hz, $f_{\text{stop1}} = 58$ Hz, and $f_{\text{stop2}} = 62$ Hz

Solution: Using the material of Section 2.5, the important values for this example and the normalized transfer function are listed below. In this case, f_{pass1} must be changed to 71.92 Hz to provide symmetry. The transfer function is shown with quadratics in factored form as shown by the (2) superscript.

$\Omega_r = 3.3$ $\quad$ $n = 2.75$ (3rd order)

$\omega_o = 376.78$ rad/sec $\quad$ $BW = 137.73$ rad/sec

$$H_{E,3}(S) = \frac{0.73880 \cdot (S \pm j6.31445)}{(S + 0.73880) \cdot (S + 0.35753 \pm j0.93390)}$$

After making the substitution of Equation 3.39 and factoring again, the following equation emerges.

$$H_{I,6}(s) = \frac{(s^2 + 141{,}960) \cdot (s \pm j387.85)^{(2)} \cdot (s \pm j366.03)^{(2)}}{(s^2 + 186.42 \cdot s + 141{,}960) \cdot (s + 16.300 \pm j322.99)^{(2)}(s + 22.126 \pm j438.42)^{(2)}}$$

After simplification, the following transfer function results.

$$H_{E,6}(s) = \frac{(s^2 + 141{,}960) \cdot (s^2 + 150{,}420) \cdot (s^2 + 133{,}980)}{(s^2 + 186.42 \cdot s + 141{,}960) \cdot (s^2 + 32.600 \cdot s + 104{,}590)(s^2 + 44.252 \cdot s + 192{,}700)}$$

We can also use FILTER to design the bandstop elliptic filter of Example 3.8. The results of this filter design are shown in Figures 3.8 and 3.9 which show the coefficients and magnitude response, respectively. Again, as in the bandpass case, a couple of the numerator coefficients are very small and can be considered zero. The passband edge frequency gain is determined to be –0.30 dB, while the stopband edge frequency gain is –56.7 dB.

```
                         Elliptic Bandstop Filter

                 Filter implementation is:      Analog
                 Filter approximation is:       Elliptic
                 Filter selectivity is:         Bandstop
                 Passband gain - lower (dB):    -0.30
                 Passband gain - upper (dB):    -0.30
                 Stopband gain         (dB):    -50.00
                 Passband freq - lower (Hz):    50.00
                 Passband freq - upper (Hz):    71.92
                 Stopband freq - lower (Hz):    58.00
                 Stopband freq - upper (Hz):    62.00

                             Order = 6
                      Overall Gain = 1.00000000e+000

       Numerator Coefficients                Denominator Coefficients
  [S^2 +       S          +      ()     ]  [S^2 +       S          +      ()     ]
  ==================================  ==================================
1 1.0  0.00000000e+00 1.41964390e+005  1.0 1.86421021e+02 1.41964390e+005
2 1.0 -4.7495764e-14 1.50423862e+005  1.0 3.26004384e+01 1.04584516e+005
3 1.0  4.48247175e-14 1.33980658e+005  1.0 4.42522615e+01 1.92704319e+005
```

Figure 3.8 Filter coefficients for Example 3.8.

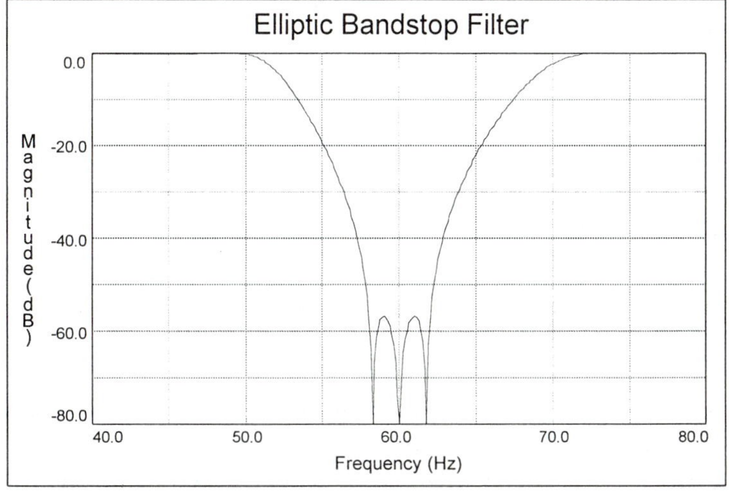

Figure 3.9 Filter magnitude response for Example 3.8.

3.5 C Code for Unnormalized Approximation Functions

We are now ready to develop the code to unnormalize the normalized lowpass coefficients which were determined in Chapter 2. We can refer to our project outline in order to determine our location in the development process. The project outline is given below where item I.B.1.c has been expanded to reveal the details of the unnormalization process. We will discuss the C code to implement each of these new items.

Analog and Digital Filter Design Project

I. Design analog or digital filter.
  A. Get filter specifications from user.
  B. Design filter according to specifications.
    1. Design analog, if designated.
      a. Calculate required order.
      b. Calculate coefficients.
      c. Unnormalize coefficients.
        (1) Determine Wo and BW.
        (2) Unnormalize lowpass, if required.
        (3) Unnormalize highpass, if required.
        (4) Unnormalize bandpass, if required.
        (5) Unnormalize bandstop, if required.
    2. Design digital FIR, if designated.
    3. Design digital IIR, if designated.
  C. Display filter parameters on screen.
  D. Save filter parameters to file.
II. Determine frequency response of filter.
III. Display magnitude and phase response on screen.

3.5.1 Unnormalize_Coefs Function

We begin with `Unnormalize_Coefs`, which is given in Listing 3.1. This function first determines the variables used for unnormalization. The unnormalization frequency `freq` for lowpass and highpass, as well as the center frequency `Wo` and the bandwidth `BW` for bandpass and bandstop cases are determined differently for the inverse Chebyshev case as compared to the other approximation methods. After these calculations, the appropriate unnormalization function is chosen based on the selectivity of the filter. Each of the specific functions called uses the `Filt_Params` structure pointer `FP` as well as the appropriate unnormalization variables. Any errors which occur in the functions are handled by the procedures discussed in Chapter 2.

```
/*=======================================================
  Unnormalize_Coefs() - converts normal lowpass coefs
                to unnormalized LP/HP/BP/BS.
  Prototype:  int Unnormalize_Coefs(Filt_Params *FP);
  Return:     error value
  Arguments:  FP - ptr to struct holding filter params
=======================================================*/
int Unnormalize_Coefs(Filt_Params *FP)
{ int     Error;    /*  error value */
  double  freq,  /*  unnormalizing freq for LP & HP  */
          BW,    /*  unnormal. bandwidth for BP & BS */
          Wo;    /*  unnormal. ctr freq for BP & BS  */

  /*  Calc freq, Wo and BW based on approx method */
  switch(FP->approx)
  { case 'B':
    case 'C':
    case 'E':
      freq = FP->wpass1;
      Wo = sqrt(FP->wpass1 * FP->wpass2);
      BW = FP->wpass2 - FP->wpass1;
      break;
    case 'I':
      freq = FP->wstop1;
      Wo = sqrt(FP->wstop1 * FP->wstop2);
      BW = FP->wstop2 - FP->wstop1;
      break;
    default:
      return ERR_FILTER;
  }
  /*  Call unnormal. function based on selectivity  */
  switch(FP->select)
  { case 'L':
      Error = Unnorm_LP_Coefs(FP,freq);
      if(Error) { return 10*Error+1;}
      break;
    case 'H':
      Error = Unnorm_HP_Coefs(FP,freq);
      if(Error) { return 10*Error+2;}
      break;
```

```
        case 'P':
          Error = Unnorm_BP_Coefs(FP,BW,Wo);
          if(Error) { return 10*Error+3;}
          break;
        case 'S':
          Error = Unnorm_BS_Coefs(FP,BW,Wo);
          if(Error) { return 10*Error+4;}
          break;
        default:
          return ERR_FILTER;
      }
      return ERR_NONE;
    }
```

Listing 3.1 Unnormalize_Coefs function.

3.5.2 Unnorm_LP_Coefs Function

Listing 3.2 contains the `Unnorm_LP_Coefs`, which handles the lowpass unnormalization. The function first determines whether there is a first-order factor by determining if the order of the approximation is odd. Remembering our technique of always placing first-order factors in the coefficient arrays first, we can safely refer to the constant term coefficients using the index of 2. (The coefficients are arranged with the s^2-term coefficient first, then the s-term coefficient, and finally the constant term coefficient. Each additional quadratic factor then follows in the same order.) The second-order factors are handled within a `for` loop that has a starting limit of `qd_start` and an ending limit of `(FD->order+1)/2`. The variable `qd_start` is set appropriately by the previous `if` statement, so that if a first-order factor has been included, the loop will start with the second quadratic. The upper limit of the `for` loop uses C integer math to properly control the number of quadratic factors to unnormalize. (Show yourself, for example, that the proper number of quadratic coefficients are unnormalized for both a fifth-order and sixth-order filter.) The coefficients within the loop are adjusted as we determined in Section 3.1. Proper indexing is accomplished by using `cf` to index individual coefficients based on `qd`, the quadratic indicator. Using this technique, any coefficient using `cf` as an index is referring to an s^2-term coefficient, while if the coefficient has an index of `cf+1`, it is an s-term coefficient, and `cf+2` will be the index for the constant term of a quadratic expression.

```
/*=======================================================
  Unnorm_LP_Coefs() - converts normal lowpass coefs to
              unnormal LP coefs at a specific freq.
  Prototype:  int Unnorm_LP_Coefs(Filt_Params *FP,
                                         double freq);
  Return:     error value
  Arguments:  FP - ptr to struct holding filter params
              freq - unnormalization frequency
=======================================================*/
int Unnorm_LP_Coefs(Filt_Params *FP,double freq)
{ int qd,cf,            /*  quad and coef number */
      qd_start;         /*  starting quad for loop */

  /*  Handle first-order, if odd; set qd_start */
  if(FP->order % 2)
  { FP->acoefs[2] *= freq;
    FP->bcoefs[2] *= freq;
    qd_start = 1;
  }
  else
  { qd_start = 0;}

  /*  Handle quadratic factors, qd indexes through
      quadratic factors, cf converts to coef number */
  for(qd = qd_start; qd < (FP->order + 1)/2; qd++)
  { cf = qd * 3;
    FP->acoefs[cf+1] *= freq;
    FP->acoefs[cf+2] *= (freq * freq);
    FP->bcoefs[cf+1] *= freq;
    FP->bcoefs[cf+2] *= (freq * freq);
  }
  return ERR_NONE;
}
```

Listing 3.2 Unnorm_LP_Coefs function.

The `Unnorm_HP_Coefs` function is very similar to the previous function and therefore will not be discussed here. That function can be found in the \FILTER\F_DESIGN.C module.

3.5.3 Unnorm_BP_Coefs Function

Listing 3.3 shown below gives the `Unnorm_BP_Coefs` function. This function and the `Unnorm_BS_Coefs` function are more complicated than the lowpass and highpass functions. One of the ways that these functions are more complicated is that the coefficient arrays must be resized to hold the larger number of coefficients for a bandpass function. Thus, early in this function, variables from the original lowpass function are stored as well as the new order of the bandpass function. Then new pointers to the larger arrays of bandpass coefficients are assigned while leaving the original pointers unchanged. Throughout this function new_num, new_den, org_num and org_den are used to identify the new and original numerator and denominator coefficients, respectively. As in previous functions, we handle the first-order factors before the second-order factors. The values assigned to the new coefficients are the same as we determined in Section 3.3.

```
/*==========================================================
  Unnorm_BP_Coefs() - converts normal lowpass coefs to
              unnormal BP coefs at a specific freq.
  Prototype:  int Unnorm_BP_Coefs(Filt_Params *FP,
                                  double BW,double Wo);
  Return:     error value
  Arguments:  FP - ptr to struct holding filter params
              BW - bandwidth for unnormalization
              Wo - center freq for unnormalization
==========================================================*/
int Unnorm_BP_Coefs(Filt_Params *FP,
                                  double BW,double Wo)
{ int      qd,ocf,ncf,qd_start,/* loop cntrs, indexes*/
           numb_coefs,          /* num coefs in array */
           org_quads,           /* orig num of quads  */
           org_order;           /* original order     */
  double   *org_num,*org_den,   /* orig num, den ptrs */
           *new_num,*new_den;   /* new num, den  ptrs */
  complex A,B,C,D,E;            /* temp cmplx vars    */

  /*  Store orig number of quads and order,
      new order will be twice original.  */
  org_order = FP->order;
  org_quads = (org_order + 1)/2;
  FP->order = org_order * 2;
  /*  For clarity, assign ptrs to temp variables  */
  org_num = FP->acoefs;
  org_den = FP->bcoefs;
  /*  Three coefs for each new quad=3*(new_order+1)/2,
      but new_order will be even, so its simplified */
```

```
numb_coefs = 3 * org_order;
/*  Allocate memory for new arrays with more coefs*/
new_num=(double *)malloc(numb_coefs*sizeof(double));
if(!new_num)  { return ERR_ALLOC;}
new_den=(double *)malloc(numb_coefs*sizeof(double));
if(!new_den)  { return ERR_ALLOC;}
/*  If org_order odd, convert first-order factor to
    quadratic, qd_start indic start pt for loop */
if(org_order % 2)
{ new_num[0] = org_num[1];
  new_num[1] = BW * org_num[2];
  new_num[2] = org_num[1] * Wo * Wo;
  new_den[0] = org_den[1];
  new_den[1] = BW * org_den[2];
  new_den[2] = org_den[1] * Wo * Wo;
  qd_start = 1;
}
else
{ qd_start = 0;}
/*  Each orig quad term will be converted to two new
    quads via complex quadratic factoring. */
for(qd = qd_start;qd < org_quads;qd++)
{ /*  ocf indexes org coefs, 3 coefs per org quad
      ncf indexes new coefs, 6 coefs per org quad
      ncf also adjusts for first-order factor */
  ocf = qd * 3;
  ncf = qd * 6 - qd_start * 3;
  /*  For numers which DON'T have s^2 or s terms. */
  if(org_num[ocf] == 0.0)
  { new_num[ncf] = 0.0;
    new_num[ncf+1] = sqrt(org_num[ocf+2]) * BW;
    new_num[ncf+2] = 0.0;
    new_num[ncf+3] = 0.0;
    new_num[ncf+4] = sqrt(org_num[ocf+2]) * BW;
    new_num[ncf+5] = 0.0;
  }
  /*  For numers which DO have s^2 and s terms. */
  else
  { /*  Convert coefs to complex, then factor */
    A = cmplx(org_num[ocf],0);
    B = cmplx(org_num[ocf+1],0);
    C = cmplx(org_num[ocf+2],0);
    cQuadratic(A,B,C,&D,&E);
```

```
         /*  Make required substitutions, factor again */
         A = cmplx(1,0);
         B = cmul(cneg(D),cmplx(BW,0));
         C = cmplx(Wo*Wo,0);
         cQuadratic(A,B,C,&D,&E);
         /*  Determine final values for new coefs. */
         new_num[ncf] = 1.0;
         new_num[ncf+1] = -2.0 * creal(D);
         new_num[ncf+2] = creal(cmul(D,cconj(D)));
         new_num[ncf+3] = 1.0;
         new_num[ncf+4] = -2.0 * creal(E);
         new_num[ncf+5] = creal(cmul(E,cconj(E)));
      }
      /*  Denoms will always have nonzero s^2 term. */
      /*  Convert coefs to complex, then factor */
      A = cmplx(org_den[ocf],0);
      B = cmplx(org_den[ocf+1],0);
      C = cmplx(org_den[ocf+2],0);
      cQuadratic(A,B,C,&D,&E);
      /*  Make required substitutions, factor again */
      A = cmplx(1,0);
      B = cmul(cneg(D),cmplx(BW,0));
      C = cmplx(Wo*Wo,0);
      cQuadratic(A,B,C,&D,&E);
      /*  Make required substitutions, factor again */
      new_den[ncf] = 1.0;
      new_den[ncf+1] = -2.0 * creal(D);
      new_den[ncf+2] = creal(cmul(D,cconj(D)));
      new_den[ncf+3] = 1.0;
      new_den[ncf+4] = -2.0 * creal(E);
      new_den[ncf+5] = creal(cmul(E,cconj(E)));
   }
   /*  Free the memory allocated to original coefs. */
   free(FP->acoefs);
   free(FP->bcoefs);
   /*  Assign the new ptrs to old array ptrs. */
   FP->acoefs = new_num;
   FP->bcoefs = new_den;
   return ERR_NONE;
}
```

Listing 3.3 Unnorm_BP_Coefs function.

Before we begin the discussion of the unnormalization of second-order factors within the `for` loop, we need to make a slight excursion into the use of complex numbers in C. The C language does not support complex numbers as a standard data type as it does `ints`, `floats`, and `doubles`. Therefore, if we are to use them in the solution of the unnormalization of bandpass and bandstop coefficients, we will have to define our own complex number definition. A complex number can easily be defined with a structure using a real and imaginary member as shown below.

```
typedef struct
{ double  re,    /*  Real part of complex number */
          im;    /*  Imag part of complex number */
}    complex;
```

Using this `complex struct` will allow us to define any variables we like as type `complex`. For example, in the variable declaration section of the function we are studying now, the variables A, B, C, D, and E are defined as `complex` as shown below.

```
complex A,B,C,D,E;              /* temp cmplx vars     */
```

Within the \FILTER\COMPLEX.C module there are a number of complex functions defined to implement standard mathematical functions for complex numbers. A list of the functions is shown below. We do not have space to study them here, but the full module is contained on the software disk included with this text.

cadd() — adds complex numbers and returns result.
cang() — returns angle (radians) of a complex number.
cconj() — returns complex conjugate of a complex number.
cdiv() — divides complex numbers and returns result.
cimag() — returns imaginary part of a complex number.
cmag() — returns the magnitude of a complex number.
cmplx() — returns complex number made from two doubles.
cmul() — multiplies complex numbers and returns result.
cneg() — returns the negative of a complex number.
cprt() — prints the value of a complex number.
cQuadratic() — factors quadratic equation with complex coefficients.
creal() — returns real part of a complex number.
csqr() — returns the square root of a complex number.
csub() — subtracts complex numbers and returns result.

As an example of one of these complex functions, cadd is shown in Listing 3.4. As indicated in the listing, cadd takes two complex numbers as arguments and adds their respective real and imaginary parts. The new complex number x is then returned to the calling function.

```
/*=========================================================
  cadd() - adds complex numbers (a+b), returns result
=========================================================*/
complex cadd(complex a,complex b)
{ complex x;
  x.re = a.re + b.re;
  x.im = a.im + b.im;
  return x;
}
```

Listing 3.4 cadd function.

Another complex function is shown in Listing 3.5 below. cQuadratic takes five arguments, all of which are complex or point to complex numbers. The first three arguments, a, b, and c, are the coefficients of a quadratic equation, while d and e are the addresses where the factors of the quadratic equation are to be stored. Each line of the function makes a partial calculation of the quadratic formula, but using complex functions. The final results are then stored where the variables d and e point.

```
/*=========================================================
  cQuadratic() - solves quadratic equation with cmplx
    coefficients. Equation form is a*x^2 + b*x + c,
    solutions will be placed in cmplx numbers d and e,
    whose addresses are sent to cQuadratic.
=========================================================*/
void cQuadratic(complex a,complex b,complex c,
                               complex *d,complex *e)
{ complex a2,ac4,sq;  /*  intermediate values */

  a2  = cmul(a,cmplx(2,0));              /*  2*a        */
  ac4  = cmul(cmul(a,c),cmplx(4,0)); /*  4*a*c         */
  sq  = csqr(csub(cmul(b,b),ac4)); /* sqrt(b*b-4*a*c)*/
  *d  = cdiv(cadd(cneg(b),sq),a2);    /*  first root  */
  *e  = cdiv(csub(cneg(b),sq),a2);    /*  second root */
}
```

Listing 3.5 cQuadratic function.

Now we can return to the discussion of the `Unnorm_BP_Coefs`. `qd_start` is again used to control the starting point of the `for` loop, and `org_quads` controls the ending point. Once we enter the loop to unnormalize the original second-order factors we first define indexing variables to control the location within the original coefficient array and the new coefficient array. The variable `ocf` controls the position within the original coefficient array by indexing three positions for each original quadratic factor. The variable `ncf` controls the position within the new coefficient array by indexing six positions for each of the original quadratics. The six positions are necessary because for each original quadratic there will be two new quadratics produced. In addition `ncf` must adjust for the unnormalized first-order factor if there is one. Therefore, `ncf` starts at 0 if the original order was even, but starts at 3 if the order was odd. Thereafter, it increments by a value of 6.

The numerator unnormalization can be of two different types. In the Butterworth and Chebyshev cases, there will be no s^2-term or s-term, while the inverse Chebyshev and elliptic approximations will have an s^2-term present for the complex zeros. Therefore an `if` statement is used to determine the appropriate method to use for a particular case. In the first case, only s-term coefficients take on the nonzero values we determined in Section 3.3. In the second case, all coefficients take on nonzero values which are dependent on a number of complex calculations.

In this second, more complicated case, we must first convert the original coefficients into complex numbers using the `cmplx` function. Then the roots of the quadratic are determined by the `cQuadratic` function and stored in the variables `D` and `E`. (We will not be dealing with the `E` root of this first quadratic because we know that it is the complex conjugate of the `D` root.) We are then ready to define another quadratic as we found in Section 3.3. The appropriate values from this quadratic are loaded into `A`, `B`, and `C`, and then `cQuadratic` is called again. The roots of this second quadratic are then stored in `D` and `E` again. Each of these roots will define one quadratic, just as one pole location in Chapter 2 was enough to determine a quadratic. For example, if $D = \alpha + j\beta$ and $E = \delta + j\lambda$, each will produce a quadratic of the forms $s^2 - 2\,\alpha\,s + (\alpha^2 + \beta^2)$ and $s^2 - 2\,\delta\,s + (\delta^2 + \lambda^2)$. These representations are mirrored in the C code. Note that the sum of squares is calculated by multiplying the root by its complex conjugate. The denominator quadratics are calculated in just the same manner as this numerator case.

Once we leave the loop, all of the new coefficients have been calculated and put in place. All we have left to do is to reassign the array pointers in the `Filt_Params` structure. We are now finished with the original coefficients from the lowpass normalized approximation, so we can free the memory allocated to them by using the `free` command. Next, we take the pointers for the new, larger arrays and put them into `FP`. So as we leave this function, the pointers in `FP` point to areas in memory which store the larger arrays of bandpass coefficients, and the areas in memory which stored the original coefficients have been freed for future use.

Although, the description of the `Unnorm_BP_Coefs` was a lengthy, complicated ordeal, we have covered all of the necessary description for `Unnorm_BS_Coefs` as well. There are really no significant differences in the bandstop unnormalization function. It can be found in the \FILTER\F_DESIGN.C module.

3.6 Displaying and Saving the Filter Parameters

We have now completed item I.B.1 on our project outline as shown below. We will need to return to items I.B.2 and 1.B.3 later in this text when we discuss the design of digital filters. Now we need to complete the analog design by displaying and saving the results as indicated in I.C and I.D.

Analog and Digital Filter Design Project

I. Design analog or digital filter.
   A. Get filter specifications from user.
   B. Design filter according to specifications.
      1. Design analog, if designated.
      2. Design digital FIR, if designated.
      3. Design digital IIR, if designated.
   C. Display filter parameters on screen.
   D. Save filter parameters to file.

II. Determine frequency response of filter.

III. Display magnitude and phase response on screen.

Actually, back in Chapter 1, the functions `Display_Filter_Params` and `Save_Filter_Params` were first mentioned in the edited main program. That section of code is shown again in Listing 3.6. Note that we first display the filter information, then obtain a suitable filename for all of the data files, and finally save the filter information to a disk file. We will discuss each of these functions in this section.

```
/*  Display both params and data to screen */
Display_Filter_Params(FP,descript);
/*  If user wants to save data, get filename */
/*  If filename is zero-length, set back to NULL */
basename = Get_FILTER_Basename();
if(!basename) { Bail_Out(
      "Error in Get_FILTER_Basename!",ERR_NULL);}
if(strlen(basename) == 0)
{ free(basename);
  basename = NULL;
}
/*  Save filter data if filename specified */
Error = Save_Filter_Params(FP,descript,basename);
if(Error) { Bail_Out(
            "Error in Save_Filter_Params!",Error);}
```

Listing 3.6 Partial listing of FILTER main function.

3.6.1 Display_Filter_Params Function

The `Display_Filter_Params` function is a rather long function and can be found in the \FILTER\F_DESIGN.C module. (The function can be viewed using a standard editor or word processor while referring to this discussion.) The first function activity is to print the variable `descript`, which holds the filter description, in the center of the screen. Then the filter characteristics are printed based on the implementation, approximation, and selectivity of the filter. Next the gain and frequency characteristics are printed with variations based on the filter's selectivity. Finally, the coefficients are printed based on the filter's implementation. In the analog case, the numerator and denominator coefficients for each quadratic factor are displayed on a single line. Throughout this function the variable `lines` is used to keep track of the number of lines that have been printed to the screen. Once the coefficients begin to be printed, this parameter is checked to make sure that no information scrolls off the screen before the user gets a chance to see it. If `lines` reaches 24, the display process stops and the user is prompted to press a key to continue. This same prompt is used at the end of the display function so that the user can view or print the screen before continuing.

3.6.2 Get_FILTER_Basename Function

We may also want to save the filter characteristics in a more permanent manner by storing them in a disk file. However, before we can save the parameters, we must get a filename from the user. This action is handled by the `Get_FILTER_Basename` function within the \FILTER\FILTER.C module, which is shown in Listing 3.7. In this function, the first requirement is to print a message to the user asking for a basename for the files which can be saved. The message indicates that if no information is to be saved, the user can simply press "Enter" and continue in the program. The types of files which can be saved are listed with their specific extensions. This entire message is required only the first time this function is called, therefore `ONCE` is defined as a `static int` to keep track of how many times this function is called. A `static` variable retains its value between function calls unlike `auto` variables which are typically used within functions. After the message display, `Get_String` is used to get the basename of the file.

```
/*======================================================
  Get_FILTER_Basename() - returns a base filename
  for all data storage files
  Prototype:  char *Get_FILTER_Basename(void);
  Return:     ptr to base filename
  Arguments:  none
=====================================================*/
char *Get_FILTER_Basename(void)
{ char *filename;          /*  temp filename ptr */
  static int ONCE = 0;     /*  indicates numb of calls */

  printf("\n Enter a filename (without extension)");
  printf(" to save the frequency response and");
  printf("\n parameters for this design. Press ");
  printf("'Enter' if no results are to be saved.");
  if(!ONCE++)
  { /*  Give user info on extensions once */
    printf("\n Specific extensions as listed below");
    printf(" will be used for data files.");
    printf("\n\n    FILE EXTENSION   ");
    printf("    FILE TYPE AND CONTENTS");
    printf("\n    .DTX & .DTB           Text and");
    printf(" binary files of params and coefs");
    printf("\n    .MAG & .MGB           Text and");
    printf(" binary files of magnitude response");
    printf("\n    .ANG & .AGB           Text and");
```

```
      printf(" binary files of phase response");
    }
    /*  Get filename */
    printf("\n");
    filename = Get_String("\n Please enter filename: ");
    if(!filename) { return NULL;}
    return filename;
}
```

Listing 3.7 `Get_FILTER_Basename` function.

After `Get_FILTER_Basename` returns the filename to the main program as shown in Listing 3.6, we check to see if `filename` has zero length. If it does, then the user did not want to save data for this run and `filename` is set back to NULL. This allows other functions which take the `filename` variable to easily check to see if a valid filename has been entered.

As indicated in the message generated by `Get_FILTER_Basename`, there are three basic file types which are saved in text and binary form. The three different files types are for the filter parameter data, the magnitude frequency response data, and the phase frequency response data. (The frequency response data will be discussed in more detail in the next chapter.) The text form is necessary for the user to review the data, while the binary form makes the transferal of data to other programs much easier.

filename.DTX — contains filter parameters in text form.
filename.DTB — contains filter parameters in binary form.
filename.MAG — contains the magnitude response in text form.
filename.MGB — contains the magnitude response in binary form.
filename.ANG — contains the phase response in text form.
filename.AGB — contains the phase response in binary form.

3.6.3 Save_Filter_Params Function

Our final function to be discussed in this chapter is item I.D of our project outline. The function `Save_Filter_Params` is too long to be included here, but it can be found in the \FILTER\F_DESIGN.C module. (The function can be viewed using a standard editor or word processor while referring to this discussion.) We first check to see if `filename` is NULL, and if it is, we return directly to the calling function with no error. If a filename is supplied, we copy the filename to a dummy string and add extensions for the binary and text files. After each of these files is opened, we are ready to write data to them. If for any reason, the files cannot be opened, we return to the calling function with an error. For most of this function we follow the same sequence as in the `Display_Filter_Params` function, except that we use `fprintf` instead of `printf` to write data to the output text file instead of the screen. In addition, variables for the binary file are prepared as we proceed through the function and `fwrite` is used to write to the binary file.

A binary file has a number of advantages over a text file in this situation. First, the binary file will be easier to read if the data is needed by another program. There is a very good chance that the filter designer will want to transfer the filter coefficients that have been determined by FILTER to another program for testing or implementation. (We will see examples of this later in the text.) Second, the coefficients and data can be stored with more accuracy than they can be displayed.

3.6.4 Binary File Format

In order to better understand the binary file structure, the format of the file is shown below. The first 4 bytes in the file will be an acronym for **A**nalog and **D**igital **F**ilter **D**esign. The next byte is a *c* to indicate that the file contains the coefficients among other parameters. The next 3 bytes are characters indicating the implementation, approximation, and selectivity types. The next 40 bytes are dedicated to the text description for the filter. The next 8 bytes are separated into 6 bytes of reserved space to allow for future indicators, and then 2 bytes are used to store the order of the filter. The filter order is placed into the binary buffer in such a manner that it could be read as an integer if desired. (The PC format for disk storage dictates that the low-order byte be written before the high-order byte for an integer. Other machines may use other methods, but will be consistent between read and write operations.) Next, 10 variables of size `double` are stored which includes the sampling frequency, the gains and frequencies in the order indicated in the listing, and the gain of the filter. And, finally, the `acoefs` and `bcoefs` variables are written as `doubles` in the order that they were stored by the program.

Header (8 bytes) — Contains the identification "ADFDc" followed by three characters indicating the implementation, approximation and selectivity types, respectively.

Description (40 bytes) — Contains the text description of the filter and is filled by nulls at the end of the description to make exactly 40 characters.

Order (8 bytes) — Contains six bytes of reserved space followed by two bytes indicating the order of the filter. The order can be read as an integer.

Specifications (10 doubles) — Contains the sampling frequency in Hertz, followed by the gains and the frequencies from the specifications, and finally the gain. The gains are specified in dB and are in the order of a_{pass1}, a_{pass2}, a_{stop1} and a_{stop2}. The frequencies are specified in Hertz and are in the order of f_{pass1}, f_{pass2}, f_{stop1}, and f_{stop2}.

A coefficients (variable) — The acoefs written as doubles.

B coefficients (variable) — The bcoefs written as doubles.

As we progress through the `Save_Filter_Params` function the proper characteristics from the first three parts of the binary file (Header, Description and Order) are stored in `buffer`. When all appropriate information has been transferred to `buffer`, it is written to the binary file using the `fwrite` command. Similarly, the data in the specification portion of the file is transferred to `dbuff` and is written to the binary file. The coefficients need not be transferred to a buffer, since we can write them to disk directly from their memory locations. The `fwrite` command only needs to know the address where the information is stored, the size of each component of information, the number of components which will be written, and the file to which they will be written. In each case of writing to the file, we check to see that the proper number of components has been written, and if not, we return an error to the calling function.

When we have reached the end of this function, we will have created both a text and binary version of all of the important information which has been determined by our filter design program. The text file will be easy to read directly, but will not have the accuracy of coefficients which we might require. The binary file will be smaller and easier to read by other programs we may wish to write, but will not find it easy to read.

3.7 Conclusion

We have now completed a large component of the project which we set out to design at the start of this text. We have completed the design of a number of analog filter types and are able to display the coefficients and data as well as save these results in a permanent form. But we can't stop here; we have yet to check the frequency response of our filter and wouldn't it be nice to see that frequency response on the screen?

Chapter 4

Analog Frequency Response Calculation and Display

In the last few chapters we have developed the necessary foundation to design a variety of analog filters. We have calculated the coefficients and are ready to implement the filter in hardware. But before we address the implementation issues in the next chapter, we need to check our design by determining the frequency response of the filter and comparing it to our design specifications. We will discuss the techniques used for frequency response determination in the first section of this chapter. Then in the next section, we will update our filter design project outline to include the calculation of frequency response and generate the C code for the frequency response determination. In the third section of this chapter, we will again expand our project outline to include the display of the frequency response on the computer screen as well as generate the C code to accomplish this. And, finally, we will discuss the completed FILTER `main` function as well as the generation of the executable program.

4.1 Calculation of Analog Frequency Response

The filter approximation function, which we determined in Chapter 3 by the calculation of the unnormalized coefficients, represents a transfer function of a linear system in the s-domain. In order to determine the frequency response of the transfer function, we must substitute $j\omega$ for each of the s-variables in that transfer function. For example, Equation 4.1 shows a transfer function with one quadratic factor, while Equation 4.2 shows the frequency response for that transfer function.

$$H(s)\big|_{s=j\omega} = \left.\frac{a_o \cdot s^2 + a_1 \cdot s + a_2}{b_o \cdot s^2 + b_1 \cdot s + b_2}\right|_{s=j\omega} \tag{4.1}$$

$$H(j\omega) = \frac{a_o \cdot (j\omega)^2 + a_1 \cdot (j\omega) + a_2}{b_o \cdot (j\omega)^2 + b_1 \cdot (j\omega) + b_2} \tag{4.2}$$

After simplification, the frequency response $H(j\omega)$ is shown as simply a frequency dependent complex number in Equation 4.3. This complex number can be represented in either rectangular form or polar form. However, when we deal with a frequency response, the polar form is the more natural form because the standard frequency response is composed of both a magnitude and phase response portion. Equation 4.4 shows the result of converting Equation 4.3 into polar form.

$$H(j\omega) = \frac{(a_2 - a_o \cdot \omega^2) + j(a_1 \cdot \omega)}{(b_2 - b_o \cdot \omega^2) + j(b_1 \cdot \omega)} \tag{4.3}$$

$$H(j\omega) = \frac{M_a \angle \tan^{-1}(P_a)}{M_b \angle \tan^{-1}(P_b)} \tag{4.4}$$

where

$$M_a = \sqrt{(a_2 - a_o\omega^2)^2 + (a_1\omega)^2}$$

$$M_b = \sqrt{(b_2 - b_o\omega^2)^2 + (b_1\omega)^2}$$

$$P_a = (a_1\omega) / (a_2 - a_o\omega^2)$$

$$P_b = (b_1\omega) / (b_2 - b_o\omega^2)$$

Of course, if the original transfer function has multiple quadratic terms, as our approximation functions do, the total frequency response is dependent on all of the quadratics. The total magnitude result will be the product of the individual magnitudes and the total phase result will be the sum of the individual phases as shown in Equation 4.5, where q represents the number of quadratic factors.

$$H_t(j\omega) = \frac{\prod_{a=1}^{q} M_a \angle \sum_{a=1}^{q} \tan^{-1}(P_a)}{\prod_{b=1}^{q} M_b \angle \sum_{b=1}^{q} \tan^{-1}(P_b)} \tag{4.5}$$

The total frequency response $H_t(j\omega)$ can then be described as shown in Equation 4.6 where the total magnitude is the numerator magnitude divided by the denominator magnitude, and the total phase angle is the denominator phase subtracted from the numerator phase.

$$H_t(j\omega) = M_t \angle \Phi_t \tag{4.6}$$

where

$$M_t = \prod_{a=1}^{q} M_a \Big/ \prod_{b=1}^{q} M_b$$

$$\Phi_t = \sum_{a=1}^{q} \tan^{-1}(P_a) - \sum_{b=1}^{q} \tan^{-1}(P_b)$$

Example 4.1 Frequency Response of a Highpass Filter

Problem: Determine the frequency response (both magnitude and phase) at the passband edge frequency of 2000 Hz and the stopband edge frequency of 800 Hz for the Chebyshev highpass filter designed in Example 3.3. Determine if the gain specifications of $a_{pass} = -1.5$ dB and $a_{stop} = -40$ dB are met. The transfer function is shown below.

$$H_{C,4}(s) = \frac{0.84140 \cdot s^2 \cdot s^2}{(s^2 + 3150.1 \cdot s + 1.6614 \cdot 10^8) \cdot (s^2 + 29{,}703 \cdot s + 6.4890 \cdot 10^8)}$$

Solution: We first make the substitution of $s = j\omega$ into the transfer function to obtain the frequency response $H(j\omega)$. Then after collecting the real and imaginary terms and arranging them, the following equation for the frequency response results.

$$H_{C,4}(j\omega) = \frac{0.8414 \cdot \omega^4}{[(1.6614 \cdot 10^8 - \omega^2) + j3150.1 \cdot \omega] \cdot [(6.4890 \cdot 10^8 - \omega^2) + j29{,}703 \cdot \omega]}$$

This frequency response equation can first be used to determine the frequency response at the stopband frequency of 800 Hz.

$$H_{C,4}(j1600\pi) = \frac{5.3713 \cdot 10^{14}}{(1.4087 \cdot 10^8 + j1.5834 \cdot 10^7) \cdot (6.2363 \cdot 10^8 + j1.4930 \cdot 10^8)}$$

$$H_{C,4}(j1600\pi) = \frac{5.3713 \cdot 10^{14}}{(1.4176 \cdot 10^{8} \angle 6.4°) \cdot (6.4126 \cdot 10^{8} \angle 13.5°)}$$

$$H_{C,4}(j1600\pi) = 5.9087 \cdot 10^{-3} \angle -19.9°$$

This indicates a gain of −44.57 dB at the stopband frequency, which exceeds the specification, and a phase shift of −19.9° or 340.1°. In the case of the passband frequency of 2000 Hz, similar calculations can be made as shown below.

$$H_{C,4}(j4000\pi) = \frac{2.0982 \cdot 10^{16}}{(8.2263 \cdot 10^{6} + j3.9585 \cdot 10^{7}) \cdot (4.9099 \cdot 10^{8} + j3.7326 \cdot 10^{8})}$$

$$H_{C,4}(j4000\pi) = \frac{2.0982 \cdot 10^{16}}{(4.0431 \cdot 10^{7} \angle 78.3°) \cdot (6.1676 \cdot 10^{8} \angle 37.2°)}$$

$$H_{C,4}(j4000\pi) = 0.8414 \angle -115.5°$$

This indicates an gain of −1.50 dB at the stopband frequency, which meets the specification, and a phase shift of −115.5° or 244.50°.

4.2 C Code for Frequency Response Calculation

We are now ready to develop the C code for determining the frequency response of an analog filter. But before we start developing code we should expand our project outline in order to effectively plan the addition of the code to this project.

4.2.1 Project Outline for Frequency Response Calculation

As shown below, item II of our project outline (initial) has been expanded to include three new items. As these items indicate, we will need to let the user specify some of the characteristics of the frequency response. And we will give the user the opportunity to save the response data to a disk file for future reference.

Analog and Digital Filter Design Project

I. Design analog or digital filter.
II. Determine frequency response of filter.
  A. Get response specifications from user.
  B. Determine frequency response.
  C. Save frequency response data to file.
III. Display magnitude and phase response on screen.

Although the three new items have helped to divide the project into more manageable tasks, there is still a need for further clarification. Each of the items can be further specified as shown below.

Analog and Digital Filter Design Project

I. Design analog or digital filter.
II. Determine frequency response of filter.
  A. Get response specifications from user.
    1. Get starting frequency.
    2. Get stopping frequency.
    3. Select dB or normal magnitude axis.
    4. Select linear or log frequency axis.
  B. Determine frequency response.
    1. Determine the total frequency response.
    2. Determine frequency response at edge frequencies.
  C. Save total frequency response to file.
    1. Save total response in text and binary form.
    2. Save edge response in text form.
III. Display magnitude and phase response on screen.

In order to properly determine the frequency response under item II.A, we need the user to specify the starting and stopping frequencies for our calculations. We also need to know whether to space the frequencies in a linear or logarithmic fashion, and whether to calculate the magnitude in decibels or not.

As indicated in II.B, we not only make frequency response calculations at the specified frequencies, but also at the critical passband and stopband edge frequencies. The calculations at the edge frequencies are necessary since the total frequency response calculation may not use the exact frequencies necessary to verify the edge frequency response. Therefore, to be certain, we will write the code necessary to make explicit calculations at the edge frequencies.

And under item II.C, we indicate that the total frequency response will be saved in both text and binary form, while the edge frequency response will be saved only in text form. The text format will allow the user to review the frequency response for specific values any time after the filter design has been completed, as long as a filename has been provided as discussed earlier. The binary format can be used by any other program which the user may want to use to analyze the data.

4.2.2 Additions to the FILTER main Function

In order to have our filter design program calculate the frequency response of our filter, we must include the necessary code in the FILTER main function. All of the functions which actually perform the frequency response calculation are contained in the \FILTER\F_RESPON.C module which has a header file of F_RESPON.H.

There is a great deal of information which must be made available to the functions in the frequency response section of the project. Therefore, we will again define a structure to hold all of the important parameters. Shown below is the `Resp_Params` structure which is used in the same manner as the `Filt_Params` structure was in the previous chapters.

```
typedef struct
{ double  *freq,          /*  array of freq values */
          *magna,         /*  array of magni values */
          *angle,         /*  array of phase values */
          *edge_mag,      /*  edge freq magnitudes */
          *edge_ang,      /*  edge freq phases */
          start_freq,     /*  start freq for resp */
          stop_freq,      /*  end freq for resp */
          gain_min;       /*  min gain to display */
  int     num_edge_freq,/*  number of edge freqs */
          decades,        /*  number of decades */
          dec_pts,        /*  number of pts/decade */
          tot_pts,        /*  total pts in resp */
          freq_axis,      /*  log or linear */
          mag_axis;       /*  log(dB) or linear */
} Resp_Params;
```

In this structure, we use pointers to store the addresses of several arrays. The variable `freq` is used for the array of frequency values to use in the response calculation, while `magna` and `angle` each refer to the magnitude response and phase response arrays. Likewise, `edge_mag` and `edge_ang` refer to the edge frequency response arrays. In addition, the starting frequency, ending frequency, and the total number of frequency points in the response are stored in the structure. The number of edge frequencies is stored since this value will vary depending on the selectivity of the filter. The characteristics of the magnitude and frequency axis are also stored in the structure as well as the number of decades and number of points per decade if the user selects a logarithmic frequency scale.

In addition to the structure definition, there are a number of defined constants included in the F_RESPON.H file as shown below. They will be used in the functions as necessary throughout the remainder of this section.

```
#define LOG      0       /*  logarithmic scale */
#define LIN      1       /*  linear scale */
#define ZERO     1E-30   /*  considered zero */
#define MAX_PTS 640      /*  max points to calc */
```

Listing 4.1 shows a section of the FILTER main function which is used to determine the frequency response of the filters which we have designed.

```
/* Allocate memory for response params struct */
RP=(Resp_Params *)calloc(1,sizeof(Resp_Params));
if(!RP)
{ Bail_Out(
        "Unable to alloc memory for RP!",ERR_ALLOC);
}
/*  Get the specs for the response from user */
Error = Get_Response_Specs(RP);
if(Error)
{ Bail_Out("Error in Get_Respons_Specs!",Error);}
/*  Calc the edge freq values and display */
Error = Calc_Edge_Resp(FP,RP);
if(Error)
{ Bail_Out("Error in Calc_Edge_Resp!",Error);}
Display_Edge_Resp(RP);
/*  Calculate the freq resp (mag and phase) */
Error = Calc_Filter_Resp(FP,RP);
if(Error)
{ Bail_Out("Error in Calc_Filter_Resp!",Error);}
/*  Save the freq response data if desired */
Error = Save_Filter_Resp(RP,descript,basename);
if(Error)
{ Bail_Out("Error in Save_Filter_Resp!",Error);}
```

Listing 4.1 Frequency response section of FILTER main function.

As indicated in the listing, we first allocate memory for the Resp_Params structure. This is necessary in order to reserve a position in memory in which to store the variables listed in the structure. The variable RP then serves as a pointer to this position in memory, and we effectively transfer all of the information within the structure simply by transferring RP.

After allocating memory for RP, we proceed through the sequence of functions as indicated in our project outline. We use Get_Response_Specs to solicit input from the user about the frequency response. We then proceed through the Calc_Edge_Resp and Display_Edge_Resp functions which calculate and display the magnitude and phase responses at the critical edge frequencies. And, finally, the Calc_Filter_Resp and Save_Filter_Resp functions are used to calculate the total frequency response and to save the response data (if the user specified a filename earlier in the program).

4.2.3 Get_Response_Specs Function

Listing 4.2 shows the Get_Response_Specs function. We first ask the user for the starting and stopping frequency for the frequency response and store the values in the Resp_Params structure. Next, the frequency resolution is set to the total number of points to be calculated in the frequency response. MAX_PTS has been defined in F_RESPON.H as 640, which is the horizontal resolution to be used in the screen display. (More discussion of the screen display functions will be given later in this chapter.)

```
/*=========================================================
  Get_Response_Specs() - get frequency response specs
  Prototype:  int Get_Response_Specs(Resp_Params *RP);
  Return:     error value
  Arguments:  RP - ptr to struct holding respon params
=========================================================*/
int Get_Response_Specs(Resp_Params *RP)
{ char    ans;        /*  answer  */
  int     resol;      /*  freq resolution (numb pts) */
  double  ratio;      /*  ratio of stop/start freqs */
  /*  Determine starting and stopping frequency */
  RP->start_freq = Get_Double
        ("\n Please specify starting frequency (Hz): ",
                                     FREQ_MIN,FREQ_MAX);
  RP->stop_freq = Get_Double
        (" Please specify stopping frequency (Hz): ",
                                RP->start_freq,FREQ_MAX);
  RP->tot_pts = MAX_PTS;
```

```
    /*  Specify the min gain to display */
    printf("\n Please specify response min gain (dB)");
    RP->gain_min = Get_Double(
          "\n (Enter 0 for linear response): ",-200,0);
    RP->gain_min = 10.0 * floor(RP->gain_min / 10.01);
    /*  Determine if magnitude should be lin or dB */
    if(RP->gain_min >= -0.001)
    { RP->mag_axis = LIN;}
    else
    { RP->mag_axis = LOG;}
    /*  Determine lin or log freq scale to be used,
        If ratio of stop/start is >= 10 use log */
    ratio=(unsigned long)(RP->stop_freq/RP->start_freq);
    if(ratio >= 10)
    { RP->freq_axis = LOG;
      RP->decades = 0;
      /* The number of decades is number of times ratio
       can be divided by 10 and still be > 2 */
      while(ratio > 2)
      { RP->decades++;
        ratio /= 10;
      }
      /*  Find integer number of points/decade,
          calc possible new tot_pts and new stop_freq to
          be integer number decades above start_freq */
      RP->dec_pts = RP->tot_pts / RP->decades;
      RP->tot_pts = RP->decades * RP->dec_pts + 1;
      RP->stop_freq=RP->start_freq*pow(10,RP->decades);
    }
    /*  If ratio not >= 10; use linear, no decades */
    else
    { RP->freq_axis = LIN;
      RP->decades = 0;
      RP->dec_pts = 0;
    }
    return ERR_NONE;
}
```

Listing 4.2 Get_Response_Specs function.

We next ask the user if the magnitude response should be calculated in dB and the value of `RP->mag_axis` is set accordingly. The frequency variable for the response can be either set to a linear or logarithmic scale and is stored in RP->freq_axis. If the ratio of the stopping frequency to starting frequency is less than 10, the linear frequency scale is selected, while if the ratio is greater than or equal to 10, the logarithmic scale is used. If a logarithmic scale is used, an integer number of decades is calculated starting at the specified starting frequency and extending to include the stopping frequency. In addition, the number of points per decade is also calculated to be an integer value, and the total number of points is adjusted if necessary. If the frequency scale is to be linear, then the number of decades and the number of points per decade are set to zero.

4.2.4 Calc_Filter_Resp Function

We will discuss the Calc_Filter_Resp function given in Listing 4.3 next since it will allow us to discuss the Calc_Edge_Resp function more efficiently later. The first activities in the Calc_Filter_Resp function are to allocate memory for the freq, magna, and angle arrays. Their pointers are stored in RP, which is used to transfer all of the information needed by the response calculation functions. Once the freq array has been allocated a space in memory, the actual values of frequency are loaded into the array based on the whether freq_axis is designated as linear (LIN) or logarithmic (LOG). In either case, freq_delta is calculated to hold the frequency increment between adjacent freq values. In the linear case, freq_delta is added to one value to get the next, while in the logarithmic case, freq_delta multiplies one value to get the next.

```
/*=======================================================
  Calc_Filter_Resp() - calcs freq response for filter
  Prototype:  int Calc_Filter_Resp(Filt_Params *FP,
                                   Resp_Params *RP);
  Return:     error value
  Arguments:  FP - ptr to struct holding filter params
              RP - ptr to struct holding respon params
=======================================================*/
int Calc_Filter_Resp(Filt_Params *FP,Resp_Params *RP)
{ int     i,          /* loop counter */
          Error;      /* error value */
  double  freq_delta; /* freq difference */

  /*  Allocate memory for freq/magna/angle arrays */
  RP->freq =
      (double *) malloc(RP->tot_pts * sizeof(double));
```

```
    if(!RP->freq)     { return ERR_ALLOC;}
    RP->magna =
         (double *) malloc(RP->tot_pts * sizeof(double));
    if(!RP->magna)   { return ERR_ALLOC;}
    RP->angle =
         (double *) malloc(RP->tot_pts * sizeof(double));
    if(!RP->angle)   { return ERR_ALLOC;}
    /*  Determine freq points for lin or log resp */
    if(RP->freq_axis == LIN)
    { freq_delta = (RP->stop_freq - RP->start_freq)
                                     / (RP->tot_pts - 1);
      RP->freq[0] = RP->start_freq;
      for(i = 1; i < RP->tot_pts; i++)
      { RP->freq[i] = RP->freq[i-1] + freq_delta;}
    }
    else
    { freq_delta = pow(10 , 1.0/RP->dec_pts);
      RP->freq[0] = RP->start_freq;
      for(i = 1 ;i < RP->tot_pts; i++)
      {   RP->freq[i] = RP->freq[i-1] * freq_delta;}
    }
    /*  Calculate frequency response based on implem */
    switch(FP->implem)
    { case 'A':
        Error = Calc_Analog_Resp(FP,RP);
        if(Error) { return 10*Error+1;}
        break;
      case 'F':
        Error = Calc_DigFIR_Resp(FP,RP);
        if(Error) { return 10*Error+2;}
        break;
      case 'I':
        Error = Calc_DigIIR_Resp(FP,RP);
        if(Error) { return 10*Error+3;}
        break;
      default:
        return ERR_FILTER;
    }
    return ERR_NONE;
}
```

Listing 4.3 Calc_Filter_Resp function.

Once the frequency values have been determined, we use the implementation type to determine which function to call in order to make the actual response calculations. The methods for determining the frequency response for analog, digital IIR, and digital FIR filters are different. In this case, we will call the `Calc_Analog_Resp` function which will be discussed next. (The other functions for determination of the frequency response of digital filters will be considered in Chapters 7 and 8.) Upon completion of the frequency response calculation, we return to the calling function. If any errors occur, they are handled in the manner we discussed in Chapter 2.

4.2.5 Calc_Analog_Resp Function

The `Calc_Analog_Resp` function is given in Listing 4.4 and performs the actual frequency response calculations for the analog filter type. After calculating the constant `rad2deg` for converting radians to degrees, the primary work of the function is performed within two nested `for` loops. The outer loop controls the frequency at which the response is being calculated, while the inner loop steps through the number of quadratics in the approximation function. As we start the calculation for a new frequency, the magnitude value (`RP->magna[f]`) is initialized to the overall gain value of the filter, while the phase value (`RP->angle[f]`) is set to 0. We then convert to radian frequencies and make the calculation of the radian frequency squared outside the quadratic loop. That saves time by not repeatedly making those calculations inside the loop where there is no change in their value.

```
/*========================================================
  Calc_Analog_Resp() - calcs response for analog filts
Prototype:  int Calc_Analog_Resp(Filt_Params *FP,
                                   Resp_Params *RP);
  Return:      error value
  Arguments:   FP - ptr to struct holding filter params
               RP - ptr to struct holding respon params
========================================================*/
int Calc_Analog_Resp(Filt_Params *FP,Resp_Params *RP)
{ int      c,f,q;          /*  loop counters */
  double   rad2deg,        /*  rad to deg conversion */
           omega,omega2,   /*  radian freq and square */
           rea,img;        /*  real and imag part */

  rad2deg = 180.0 / PI;  /*  set rad2deg */
  /*  Loop through each of the frequencies */
```

```
    for(f = 0 ;f < RP->tot_pts; f++)
    { /*  Initialize magna and angle */
      RP->magna[f] = FP->gain;
      RP->angle[f] = 0.0;
      /*  Pre calc omega and omega squared */
      omega = PI2 * RP->freq[f];
      omega2 = omega * omega;
      /*  Loop through coefs for each quadratic */
      for(q = 0 ;q < (FP->order+1)/2; q++)
      { /*  c is coef index = 3 * quad index */
        c = q * 3;
        /* Numerator values */
        rea = FP->acoefs[c+2] - FP->acoefs[c] * omega2;
        img = FP->acoefs[c+1] * omega;
        RP->magna[f] *= sqrt(rea*rea + img*img);
        RP->angle[f] += atan2(img,rea);
        /* Denominator values */
        rea = FP->bcoefs[c+2] - FP->bcoefs[c] * omega2;
        img = FP->bcoefs[c+1] * omega;
        RP->magna[f] /= sqrt(rea*rea + img*img);
        RP->angle[f] -= atan2(img,rea);
      }
      /* Convert to degrees */
      RP->angle[f] *= rad2deg;
    }
    /*  Convert magnitude response to dB if indicated */
    if(RP->mag_axis == LOG)
    { for(f = 0 ;f < RP->tot_pts; f++)
      { /* Handle very small numbers */
        if(RP->magna[f] < ZERO)
        { RP->magna[f] = ZERO;}
        RP->magna[f] = 20 * log10(RP->magna[f]);
      }
    }
    return ERR_NONE;
}
```

Listing 4.4 Calc_Analog_Resp function.

Once we reach the inner quadratic loop which is controlled by the variable q, we define a value c which is based on q to help access the individual coefficients of each quadratic. We first handle the numerator portion of the quadratic by determining the real and imaginary parts of the complex number. Then the total magnitude is multiplied by the magnitude of the complex number, and the total angle in incremented by the angle of the complex number. A similar process is performed for the denominator portion of the quadratic except that the total magnitude is divided by the denominator's magnitude, and the denominator's phase is subtracted from the total phase. This inner loop process is repeated for all of the quadratic terms, and then the phase value is converted to degrees outside the quadratic loop.

If the magnitude of the response was specified to be in decibels, each of the magnitude values is converted to decibels before leaving the function. Since the logarithm of zero is undefined, an artificial value `ZERO` is defined in F_RESPON.H so that no math error will be produced by the compiler if values become too small. Since we are using `doubles` to describe the magnitude values, a value of 1E-30 was chosen for `ZERO` which is still within the limits of expression for doubles. This value will produce an gain value of –600 dB, which is well beyond the normal levels of gain we expect in filter design, so our definition of `ZERO` should have no effect on the normal operation of the program.

4.2.6 Edge Response Functions

In this section we discuss the three functions which deal directly with the passband and stopband edge frequencies. Each of these functions can be found in the \FILTER\F_RESPON.C module on the software disk provided with this text. The `Calc_Edge_Resp` calculates the frequency response at the passband and stopband edge frequencies. This function is necessary in addition to the normal frequency response calculation function `Calc_Filter_Resp` since the normal frequency response calculations may not include frequency points at exactly the passband and stopband edge frequencies. The function is very similar to the `Calc_Filter_Resp` function. It makes use of the format of the `Calc_Filter_Resp` function by creating a `Resp_Params` structure pointer `ER` for the edge frequency response. The structure to which `ER` points is then loaded with the array of frequencies which includes only the passband and stopband edge frequencies of the filter. There will be two critical frequencies for lowpass and highpass filters and four critical frequencies for bandpass and bandstop filters. In either case, space for the frequency arrays, magnitude arrays, and phase arrays are allocated as necessary. The values of frequency are available from the `Filt_Params` structure and are converted back to Hertz before being loaded into the `freq` array.

Once the frequency array has been loaded, the proper response calculation function can be called just as in the Calc_Filter_Resp function. Upon return from the response calculation function, the pointers which hold the addresses of the magnitude and phase values for the edge frequencies are transferred to the RP structure and the ER structure is eliminated from memory. Note that this does not free the ER->magna and ER->angle pointers which have been stored in the RP structure.

The Display_Edge_Resp function allows the magnitude and phase values at the critical edge frequencies to be displayed on the screen for the user's view. This allows the designer instant evaluation of the filter at the most important frequencies. The complete response will also be calculated and displayed in graphical form, but it is the numerical values of the magnitude and phase at these edge frequencies which clearly indicate the successful design of the filter.

The Save_Filter_Resp function provides the means for saving the total frequency response data as well as the edge frequency response data. The first activity is to determine if the user actually wants to save the data. This is accomplished by testing the basename variable which is one of the arguments which is transferred to the function. If basename is NULL, then no filename was entered when requested, and therefore no data will be saved. On the other hand, if basename has a nonzero value, then it contains the filename to be used to save data. Four different filenames are constructed by adding the appropriate extension. Each of these four different files is opened at various times within the function using the fopen command. Both the magnitude and phase responses will be saved with each being saved in text and binary form.

The description of the filter is written to the text file first. Then the edge frequency magnitude data is written, followed by the complete magnitude response data. The process is then repeated for the phase information. In each file write operation, a check is made to verify the completion of the task. In each case, the file is closed after all operations have been completed.

The binary files for the magnitude and phase responses have the format as shown below. The header and description information are stored in a buffer prior to being written to the magnitude and phase files like the method used in Save_Filter_Params in Chapter 3. The frequency magnitude and phase data are written directly to the data files from memory. Again, verification of the file write procedures is made, and all files are closed prior to leaving the function.

Header (8 bytes) — Contains the identification 'ADFDm∅' for magnitude files and 'ADFDp∅' for phase files. The next two bytes, which can be read as a binary integer give the number of points in the frequency response. (The '∅' symbol represents a space.)

Description (40 bytes) — Contains the text description of the filter and is filled by nulls at the end of the description to make exactly 40 characters.

Frequency data (variable) — Contains the number of frequency values as indicated in the header with size of double.

Magnitude or phase data (variable) — Contains the number of magnitude or phase values as indicated in the header with size of double.

4.3 C Code for Frequency Response Display

We are now ready to outline the procedures necessary to display the magnitude and phase responses on the computer screen. Any programming project involving graphics adds an extra dimension to the final result and to the challenge of the project. This project is no exception. Every computer type seems to have a unique way of producing graphics. Therefore, the code in this section is unique to a PC computer system and will not work on other machines directly. Even different compiler manufacturers use different function names to accomplish similar graphics operations. However, as outlined in further detail later in this section, the code in this section has been developed in a manner to ease the implementation problems. First, three separate modules of graphics code have been developed to directly support popular PC compilers. And, second, all graphics functions have been written with generic names in the main code. Therefore, if implementation is necessary on other platforms, only those generic graphics functions must be implemented and assembled into a separate module for linking. All other code used in the various projects discussed in this text should be fairly portable.

4.3.1 Differences Between Text and Graphics Mode

Up to this point our program output has been using the screen in text mode which can display 25 lines of 80 columns each on the screen. This is a very efficient mode of operation with each of the 2000 positions on the screen being able to display 1 of over 200 characters in a variety of colors. However, when it comes to producing a graph of a frequency response, the text mode is not able to provide the necessary detail that we would like. Luckily, most of today's modern personal computer systems provide for graphics modes of operation which can do a very nice job of displaying a variety of graphical results.

In order to use the graphics capabilities of a computer system, the computer must be connected to a color monitor which can display the required resolution and must have a video graphics array (VGA) adapter to drive the color monitor. Most PC computer systems sold in the last few years have the necessary hardware to make use of the graphics functions we will be using. However, computer systems with EGA or CGA adapters will not be able to use the graphics software, and users with monochrome monitors will not be able to display the colors which will be used. We will be using one of the most basic graphics modes available on the VGA adapter for our project. Sometimes referred to as mode 18 (or in hexadecimal, 0x12), this graphics mode provides a screen layout with 640 picture elements (pixels) across and 480 pixels down the screen. This gives us over 300,000 pixels which can take on any one of 16 different colors. Each of these pixels can be individually controlled, but with all of this potential comes some additional responsibility.

In addition to the hardware concerns, there are a number of software issues to address when using graphics programming techniques. Since most computer systems start operating in text mode, special commands must be given to the system in order to switch to graphics mode. Text and graphics modes do not normally exist on the screen at the same time. Therefore, once we switch our system to graphics, all operations must be handled by special graphics functions.

One of the more difficult operations to perform in graphics mode is actually one of the simplest in text mode. Placing text on the screen in text mode is a very simple matter which can be handled by a variety of C functions. However, in graphics mode the placement of text on the screen becomes an ordeal of placing a number of colored dots on the screen to resemble a character. Each character has to be made up of a different pattern of dots. Luckily, most compilers provide a variety of fonts which have already determined the patterns to use for text characters. Actually, some compilers allow their fonts to be expanded and rotated to produce various sizes of text in various orientations which is not possible in text mode. Although the graphics mode does provide us with these additional features, we still must perform a number of steps to select particular fonts and register them for use. In addition, special output functions are required to display the text since standard C functions will not perform correctly in graphics mode.

In addition to the text which is usually a part of any graphical screen display, we will also be presenting the frequency response as a graphical plot.

Not only must the data be accurately plotted, but the plot area must be properly scaled with accurate division lines drawn for both horizontal and vertical axes. Luckily, compilers offering graphical options often have functions available to move about the screen and to draw lines and figures. In order to help visualize the display, a basic view of the screen we will be generating is shown in Figure 4.1. Notice the position of the global title as well as the titles and labels for the x and y axes. The screen area extends from the upper right coordinate (0,0) to the lower left coordinate (639,479). The plot area is defined from the upper left coordinate (*ulx,uly*) to the lower right coordinate (*lrx,lry*). The only part which is missing from this display is the plot of the data which will be contained within the plot area.

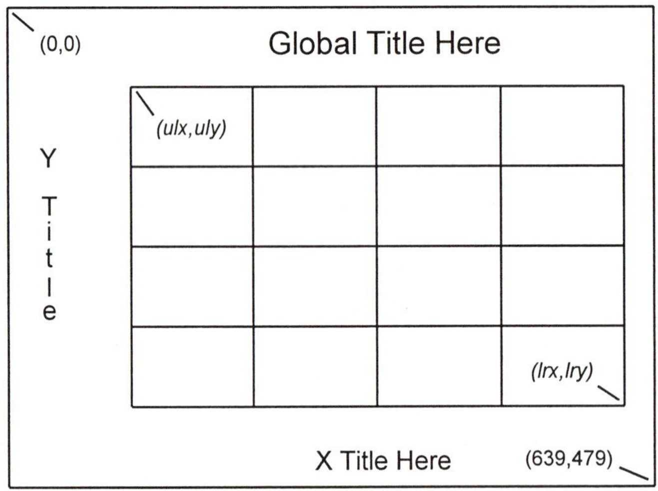

Figure 4.1 Representation of computer screen.

4.3.2 Project Outline for Frequency Response Display

Our project outline as shown below expands item III for clarification. Here we see the fundamental steps required to present a graphical representation of the frequency response on the screen. As indicated, the desired graphics mode and text font must be selected prior to the actual screen display of the frequency response information. Then after the presentation has been completed, the system must be set back to its original state.

Analog and Digital Filter Design Project

I. Design analog or digital filter.
II. Determine frequency response of filter.
III. Display magnitude and phase response on screen.
   A. Set the desired graphics mode.
   B. Select and register the desired font.
   C. Define the screen characteristics.
   D. Display the magnitude and phase responses.
   E. Remove the registered fonts.
   F. Reset the system to text mode.

Even though the expansion of item III has helped clarify the steps required in the proper operation of the graphics system, items III.C and III.D require a bit more discussion. These items have been further expanded as shown below. The screen display requirements for this project can be divided into three major areas. First, all of the text to be placed on the screen must be properly defined. This includes not only the major titles of the screen, but also the labels on the x and y axes. Second, the screen area where the plot will be made must be properly dimensioned for linear or logarithmic displays. And, finally, the plot of the data itself, whether it is the magnitude or phase response, must be properly defined. After each of these major areas has been defined, the respective data must be displayed.

Analog and Digital Filter Design Project

I. Design analog or digital filter.
II. Determine frequency response of filter.
III. Display magnitude and phase response on screen.
   A. Set the desired graphics mode.
   B. Select and register the desired font.
   C. Define the screen characteristics.
      1. Define all text items for the display.
      2. Define all aspects of the plot area.
      3. Define all aspects of the data to be plotted.
   D. Display the magnitude and phase responses.
      1. Display all text items on the screen.
      2. Display all aspects of the plot area.
      3. Display the data within the plot area.
   E. Remove the registered fonts.
   F. Reset the system to text mode.

4.3.3 Data Variables for the Screen Display

As has become commonplace with this project, the variables used to transfer information from one screen function to another will be stored in a structure. In this case, the structure is one of the type `Scrn_Params` as defined below.

```
typedef struct
{ title     *G_Title, /* global title */
            *X_Title, /* x axis title */
            *Y_Title; /* y axis title */
  label     *X_Label, /* labels for x axis */
            *Y_Label; /* labels for y axis */
  plot_area *Box;     /* plot area layout */
  plot_data *Data;    /* data to be plotted */
} Scrn_Params;
```

Within this structure we see that there are a variety of variable types which are not standard C variables. These variables (`title`, `label`, `plot_area`, and `plot_data`) are all additional structures which will be defined soon. However, before we move to the definition of these other structures, note that the primary structure, `Scrn_Params` contains the information about the global title, the x-axis title and labels, the y-axis title and labels, the plot area and the plot data. Each of these items is stored as a pointer to the appropriate type of structure.

The structure of type `title` is shown below. The variables included in the structure are the text, the alignment (left, center, right), the position, the color, and the size of the title. More discussion of the possible values for these variables will be given in the C code section.

```
typedef struct
{ char  *text,          /* text & size of title */
         align;         /* alignment (L,C,R) of title */
  int    xpos,ypos,     /* position of title */
         color,size;    /* color/size of title text */
} title;
```

The structure of type `label` is shown next. As indicated in this structure, we use an array of text strings to store the set of labels which extend along either the x or y axis. The x and y positions and alignment of the labels are also stored. The color, size, and number of labels are also part of the structure.

```
typedef struct
{ char  **text,          /* array of text labels */
        align;           /* alignment of labels */
  int   *xpos,*ypos,     /* array of label positions */
        color,size,      /* color and size of labels */
        number;          /* number of labels */
} label;
```

The plot_area structure is shown below and includes a specification of the upper left and lower right corners of the plot rectangle as well as the color and thickness of the line boxing the area. In addition, the number and color of the major and minor divisions on the x and y axes are also stored in the structure. Also the status of the x and y axis as using linear or logarithmic variations is also stored.

```
typedef struct
{ int ulx,uly,       /* upper left corner of plot */
      lrx,lry,       /* lower right corner of plot */
      thickness,     /* thickness of line boxing plot */
      color,         /* color of line boxing plot */
      X_Maj_Div,     /* number of X major divisions */
      Y_Maj_Div,     /* number of Y major divisions */
      X_Min_Div,     /* number of X minor divisions */
      Y_Min_Div,     /* number of Y minor divisions */
      Maj_Div_Color,   /* major division color */
      Min_Div_Color,   /* minor division color */
      X_LL_Type,     /* X scale log or lin indicator */
      Y_LL_Type;     /* Y scale log or lin indicator */
}   plot_area;
```

In addition, the plot_data structure is shown below and includes arrays for the x-axis and y-axis data. Also, the total number of points to be plotted is given as well as the maximum and minimum value of y data to be plotted. These later two items will limit the display to the bounded plot area.

```
typedef struct
{ double  *x_data,   /* array of x data to plot */
          *y_data,   /* array of y data to plot */
          y_max,     /* max value of y data to plot */
          y_min;     /* min value of y data to plot */
  int     tot_pts;   /* total number of points */
} plot_data;
```

Also contained in the F_SCREEN.H include file are a number of defined constants used throughout the screen display functions. Their values are shown below, and their usage will be discussed during the explanation of the function operation.

```
/*  #DEFINES:                          */
#define Black      0          #define Blue          1
#define Green      2          #define Cyan          3
#define Red        4          #define Magenta       5
#define Brown      6          #define White         7
#define LBlack     8          #define LBlue         9
#define LGreen     10         #define LCyan         11
#define LRed       12         #define LMagenta      13
#define Yellow     14         #define LWhite        15
#define MAG        0          #define ANG           1
#define LIN        1          #define LOG           0
#define HOR        0          #define VER           1
#define ULX        99         #define ULY           59
#define LRX        579        #define LRY           419
#define MAX_WIDTH 640         #define MAX_HEIGHT 480
#define LRG        2          #define MED           1
#define SML        0
```

4.3.4 Primary Program Components for Screen Display

As shown below, the portion of the FILTER main function which is associated with the display of the screen data is minimal. We first allocate memory for the `Scrn_Params` structure and call the pointer returned `SP`. We then call the `Display_Screen_Data` function which has as arguments the pointers to the `Resp_Params` structure, the `Scrn_Params` structure, the text description of the filter, and the minimum gain to be displayed on the screen.

```
/*  Alloc memory for Scrn_Params structure */
SP = (Scrn_Params *)calloc(1,sizeof(Scrn_Params));
if(!SP) { Bail_Out(
          "Unable to allocate mem for SP!",ERR_ALLOC);}
Error = Display_Screen_Data(RP,SP,
                                  descript,RP->gain_min);
if(Error) { Bail_Out(
               "Error in Display_Screen_Data!",Error);}
```

The entire job of displaying the frequency response data on the screen is handled by the `Display_Screen_Data` function, which is shown in Listing

4.5. The first order of business in this function is to set the graphics mode to the 16-color VGA mode by using `SetVGA16Color` function. Next, the proper font is selected using the `RegisterFont` function, which takes as an argument the complete pathname to the directory where the font is stored. If any error occurs when calling either of these functions, the normal error handling procedure is used except that the video mode is first set back to text mode using `SetTextMode`. This will be the case throughout this function, since leaving the video mode in graphics mode when exiting a program is considered poor programming practice.

```
/*=====================================================
  Display_Screen_Data() - displays all graphics to
  screen, including text, boxes and data
  Prototype:  int Display_Screen_Data(Resp_Params *RP,
    Scrn_Params *SP,char *descript,double Gain_Min);
  Return:     error value
  Arguments:  RP - ptr to struct holding respon params
              SP - ptr to struct holding screen params
              descript - global title
              Gain_Min - min gain to show on plot
=====================================================*/
int Display_Screen_Data(Resp_Params *RP,
      Scrn_Params *SP,char *descript,double Gain_Min)
{ char  ch;      /*  return value */
  int   Error;   /*  error value */

  /*  Set graphics, register font and clear screen */
  Error = SetVGA16Color();
  if(Error) { SetTextMode(); return (10*Error + 1);}
  Error = RegisterFont(FONT_PATH);
  if(Error) { SetTextMode(); return (10*Error + 2);}
  ClrScrn();
  /*  Load associated structures with MAG info */
  Error = Load_Plot_Data(RP,SP,MAG,Gain_Min);
  if(Error) { SetTextMode(); return (10*Error + 3);}
  Error = Load_Plot_Area(RP,SP,MAG);
  if(Error) { SetTextMode(); return (10*Error + 4);}
  Error = Load_Titles(RP,SP,MAG,descript);
  if(Error) { SetTextMode(); return (10*Error + 5);}
  Error = Load_Labels(RP,SP,MAG);
  if(Error) { SetTextMode(); return (10*Error + 6);}
  /*  Show text, box and data for MAG */
```

```
Error = Show_Text(SP);
if(Error) { SetTextMode(); return (10*Error + 7);}
Error = Show_Box(SP->Box);
if(Error) { SetTextMode(); return (10*Error + 8);}
SetColor(Yellow);
/*  Allow user to press 'P' for print if a hard copy
   of graph is desired. Other keystroke will proceed
   with program. If 'P' is pressed, color is set to
   LWhite which will print better than Yellow. */
while(1)
{ Show_Data(SP,MAG);
   ch = (char) getch();
   if((ch == 'P') || (ch == 'p'))
   { SetColor(LWhite);}
   else
   { break;}
}
ClrScrn();
/*  Now load associated structures with ANG info */
Error = Load_Plot_Data(RP,SP,ANG,Gain_Min);
if(Error) { SetTextMode(); return (10*Error + 3);}
Error = Load_Plot_Area(RP,SP,ANG);
if(Error) { SetTextMode(); return (10*Error + 4);}
Error = Load_Titles(RP,SP,ANG,descript);
if(Error) { SetTextMode(); return (10*Error + 5);}
Error = Load_Labels(RP,SP,ANG);
if(Error) { SetTextMode(); return (10*Error + 6);}
/*  Show text, box and data for ANG */
Error = Show_Text(SP);
if(Error) { SetTextMode(); return (10*Error + 7);}
Error = Show_Box(SP->Box);
if(Error) { SetTextMode(); return (10*Error + 8);}
SetColor(Yellow);
/*  Allow user to press 'P' for print if a hard copy
   of graph is desired. Other keystroke will proceed
   with program. If 'P' is pressed, color is set to
   LWhite which will print better than Yellow. */
while(1)
{ Show_Data(SP,ANG);
   ch = (char) getch();
   if((ch == 'P') || (ch == 'p'))
   { SetColor(LWhite);}
   else
```

```
        { break;}
    }
    /*  Set back to text mode and return */
    UnRegisterFont();
    SetTextMode();
    return ERR_NONE;
}
```

Listing 4.5 Display_Screen_Data function.

The `SetVGA16Color`, `RegisterFont`, and `SetTextMode` functions all carry out operations which are compiler dependent in their implementation. These functions, as well as others we will encounter along the way, have to be written using the specific functions supplied by the compiler manufacturer. Therefore, there are three special modules included with this text to supply at least three options for programmers. All of these modules will be discussed in Section 4.4.1.

Prior to beginning the real work of the function, we clear the graphics screen using a compiler-dependent function `ClrScrn`. Then the `plot_data` structure of `Scrn_Params` is loaded with appropriate values for display of the magnitude response by the `Load_Plot_Data` function. The `MAG` indicator is defined in the F_SCREEN.H include file as well as the `ANG` indicator to specify which portion of the frequency response the function is supposed to process. Next, the `Load_Plot_Area` function is used to load the `plot_area` structure of the `Scrn_Params` structure, and this is followed by loading all of the `label` and `title` structures used by `Scrn_Params`.

Once all of the structures have been loaded with the appropriate values, we can proceed to display the text, the plot area, and the data by calling the `Show_Text`, `Show_Box` and `Show_Data` functions, respectively. Once the data has been displayed on the screen the user can get a hardcopy of the screen display by pressing the "Print Screen" button on the keyboard. In order for the graphics display to be successfully printed, the user must take care of two details. First, the user must run the DOS program GRAPHICS.COM prior to running the FILTER program. (GRAPHICS.COM specifies the printer attached to the computer system so that the graphics screen can be printed properly. Of course, the printer must be able to print graphics.) Second, when the display to be printed appears on the screen, the user must press the "P" key before the "Print Screen" key. The reason for this two key sequence is as follows. The data plot is initially made using the color yellow, which shows up nicely on a black screen. (This color is set using the SetColor function, and can be changed to any other color defined in the

F_SCREEN.H file.) However, the color yellow does not show up well on a screen print so a technique to allow a better printout was developed. The technique used is to display the graph again using a color which will show up well before printing out the graph. This is accomplished by the `while` loop which includes the `getch` function. If the user presses "P" while the graph is displayed, the graph will be redisplayed using the `LWhite` color which shows up very nicely on a screen print. After the screen has been redisplayed, the "Print Screen" key can be used to print the plot.

If any other key than "P" is pressed, the program continues by clearing the screen, and the process is repeated for the phase portion of the frequency response. One addition in the phase display section is the use of the function `Limit_Data`. This function limits all angles to values within the primary range of –180 to +180 degrees. This is a more typical way of viewing a phase response which will often become quite large. This method simply adds or subtracts 360 degrees whenever the phase angle goes out of the ±180 degree range. At the completion of the screen display, the font is unregistered, and the video mode is returned to text mode and execution is returned to the calling function.

4.3.5 Functions for Displaying the Plot Data

Before we can display the frequency response data on the screen, we must set up the variables in the `plot_data` structure within the `Scrn_Params` structure. Therefore, the `Load_Plot_Data` function as shown in Listing 4.6 is used to prepare the structure as necessary. In this function, the first activity is to allocate memory for the `plot_data` structure and store the pointer to the structure in `SP->Data`. Within the `plot_data` structure, the `x_data` pointer is the same as the frequency information pointer stored in `RP->freq`. Therefore, this pointer is copied to the `x_data` pointer. If the `ma_type` variable indicates that the magnitude response data is being loaded, then the `y_data` pointer is set to `RP->magna` where otherwise it is set to `RP->angle`. The maximum and minimum values of the magnitude are set according to whether a linear or logarithm scale is to be used. The phase minimum and maximum values are set to ±180 degrees. The number of total points used in the frequency response is specified before leaving the function.

The copying of the `RP->freq`, `RP->magna` and `RP->angle` pointers can be a dangerous practice leading to unforeseen problems. The problems can result because we now have two pointers pointing to the same information in memory. If we get sloppy, one of these pointers could be used to free the

```
/*=====================================================
  Load_Plot_Data() - loads plot_data structure
  Prototype:  int Load_Plot_Data(Resp_Params *RP,
        Scrn_Params *SP,int ma_type,double Gain_Min);
  Return:     error value
  Arguments:  RP - ptr to struct holding respon params
              SP - ptr to struct holding screen params
              ma_type - mag or ang indicator
              Gain_Min - min gain to show on plot
=====================================================*/
int Load_Plot_Data(Resp_Params *RP,Scrn_Params *SP,
                          int ma_type,double Gain_Min)
{ /*  Create plot_data structure for Data */
  SP->Data = (plot_data *) calloc(1,sizeof(plot_data));
  if(!SP->Data) { return ERR_ALLOC;}
  /*  X data is freq, Y data is magna or angle
      ymax and ymin are set based on ma_type */
  SP->Data->x_data = RP->freq;
  if(ma_type == MAG)
  { SP->Data->y_data = RP->magna;
    if(RP->mag_axis == LOG)
    { SP->Data->y_max = 0.0;
      SP->Data->y_min = Gain_Max;
    }
    else if(RP->mag_axis == LIN)
    { SP->Data->y_max = 1.0;
      SP->Data->y_min = 0.0;
    }
    else { return ERR_VALUE;}
  }
  else if(ma_type == ANG)
  { SP->Data->y_data = RP->angle;
    SP->Data->y_max =  180.0;
    SP->Data->y_min = -180.0;
  }
  else
  { return ERR_VALUE;}
  /*  Set tot_pts and return */
  SP->Data->tot_pts = RP->tot_pts;
  return ERR_NONE;
}
```

Listing 4.6 Load_Plot_Data function.

memory and then at a later time, the other pointer could be used to try to access the information again. This access may provide correct information if no other data has overwritten the freed memory, but there is no guarantee that the memory will hold the expected data.

An alternative to the copying of the pointer would be to allocate new memory and copy the information to the new memory location, thereby providing two independent pointers to independent data. The disadvantage of this technique is that we are taking time to allocate new memory and copy information, and we are using a large amount of new memory. Because we don't want to waste time or memory, and because the normal operation of this program will not use the `Resp_Params` and `Scrn_Params` structures independently, we will simply copy the pointer and issue warning comments when dealing with the `RP->freq`, `RP->magna`, and `RP->angle` pointers. Once the `plot_data` structure has been loaded with the appropriate variables and information, the frequency response data can be displayed on the screen by `Show_Data` as given in Listing 4.7. In order to simplify the calculations to be made, new local variables are defined to hold the somewhat complicated variables using multiple pointer notation. Additional floating point constants `incr` and `mult` are calculated and the x and y positions of the data points are then calculated. The graphing procedure is handled primarily by the `MoveTo` and `LineTo` functions. We begin by moving to the starting point on the graph. Then each succeeding line segment of the graph is drawing using the `LineTo` function, which draws a line from the current point to the new point designated in the arguments. The `LineTo` function also automatically resets the current point on the screen to the ending point of the line. So once the process gets started, the `LineTo` function is continually called using new values for the ending point of the line. Since the range of the y data values may exceed the limits indicated for display, before each value of the of the y position is used, it is tested to verify that it is within the valid range.

The calculation of the y position of the points on the screen is best understood by equating two ratios. These ratios are formed by dividing the partial distance that the data point is from the upper y limit by the total distance between the upper and lower y limits. The first ratio is based on the screen dimensions in pixels and is shown in Equation 4.7.

$$\frac{ypos - uly}{lry - uly} = \frac{\text{partial}}{\text{total}} \tag{4.7}$$

The second ratio is based on actual y data values as shown in Equation 4.8.

$$\frac{ymax - data[i]}{ymax - ymin} = \frac{\text{partial}}{\text{total}} \tag{4.8}$$

When these two ratios are equated, and solved for `ypos`, Equation 4.9 results. A portion of this equation is constant and can be precalculated and stored as the variable `mult` as shown in Equation 4.10.

$$ypos = \frac{(lry - uly)}{(ymax - ymin)} \cdot (ymax - data[i]) + uly \tag{4.9}$$

$$ypos = mult \cdot (ymax - data[i]) + uly \tag{4.10}$$

Although this produces the basic equation, there are some additional details required to convert from floating point to integer values as accurately as possible. The technique usually used is to add a value of 0.5 to the floating point result before truncating it to an integer. For example, if the floating point value is equal to 111.6, the addition will make the value 112.1 before it is truncated to 112 in integer form.

The calculation of the x position of each point is much simpler than for the y position. All that is necessary is to equally space the points along the x dimension of the plot area. The linear or logarithmic scale which will be drawn on the plot area will take care of setting the correct scale for the frequencies.

4.3.6 Functions for Displaying the Plot Area and Graphics Text

The `Load_Plot_Area` function as well as the other functions to display the plot area for the frequency response are included in the \FILTER\F_SCREEN.C module. These functions draw the frequency response box on the screen as well as draw the major and minor division for the graph. Color, thickness, and position are all determined within these functions. The precise placement of divisions depends on whether magnitude or phase plots are being shown, and whether the frequency scale is linear or logarithmic. The `Calc_Maj_Pos` and `Calc_Min_Pos` functions are responsible for these calculations.

```
/*=========================================================
  Show_Data() - displays the graphs and legend
  Prototype:  void Show_Data(Scrn_Params *SP,
                                          int ma_type);
  Return:     error value
  Arguments:  SP - ptr to struct holding screen params
              ma_type - mag or ang indicator
=========================================================*/
void Show_Data(Scrn_Params *SP,int ma_type)
{ int     xpos,ypos,    /* x and y positions */
          ulx,uly,      /* upper left x and y */
          lrx,lry,      /* lower right x and y */
          tpts,i;       /* total pts & loop counter */
  double  *data,        /* ptr to data */
          incr,mult,    /* increment and multiplier */
          ymax,ymin;    /* max and min values of y */
  /*  Limit data if ANG */
  if(ma_type == ANG)
  { Limit_Data(SP->Data,180);}
  /*  Simplify variables by removing pointers */
  data = SP->Data->y_data;  tpts = SP->Data->tot_pts;
  ymax = SP->Data->y_max;   ymin = SP->Data->y_min;
  lrx = SP->Box->lrx;       lry = SP->Box->lry;
  ulx = SP->Box->ulx;       uly = SP->Box->uly;
  /*  Determine increment and multiplier */
  incr = (double)(lrx - ulx) / (double)(tpts - 1);
  mult = (double)(lry - uly) / (ymax - ymin);
  /*  Set starting point on left side of plot */
  xpos = ulx;
  ypos = (int)(0.5 + (ymax-data[0]) * mult) + uly;
  if(ypos > lry)  { ypos = lry;}
  if(ypos < uly)  { ypos = uly;}
  MoveTo(xpos,ypos);
  /*  Plot graph from one point to another */
  for(i = 1; i < tpts; i++)
  { xpos = ulx + (int)(incr * (double)i + 0.5);
    ypos = (int)(0.5 + (ymax-data[i]) * mult) + uly;
    if(ypos > lry)  { ypos = lry;}
    if(ypos < uly)  { ypos = uly;}
    LineTo(xpos,ypos);
} }
```

Listing 4.7 Show_Data function.

There are a number of functions used to enable the display of text on the graphics screen. These are also included in the \FILTER\F_SCREEN.C module. The primary work of initializing the text to be displayed is handled by the `Load_Titles` and `Load_Labels` functions, while the display of the text is handled by the `Show_Text`, which in turn calls the `Show_Title` and `Show_Label` functions.

4.4 Compiling the FILTER Program

We have now discussed the complete project outline for the filter design program. Although we still have to fill in the details of the digital filter design functions, these will be comparable to the analog filter design methodology. The calculation of the digital filter frequency response will be handled by different functions which will be discussed in later chapters, but the display routines will not change. Therefore, it seems a good time to discuss the methods needed to compile the separate modules of this project and to link them together into an executable program. However, before we study those details, we need to briefly discuss the compiler-specific graphics functions.

4.4.1 Compiler-specific Graphics Functions

As mentioned in the previous section, many of the graphics functions are compiler-specific and must be handled in a special way. In our screen display function in the F_SCREEN.C module, all graphics functions have been written using generic function names with all compiler-specific functions to implement the generic functions stored in their own separate module. The \FILTER\B_GRAPHX.C module and the accompanying B_GRAPHX.H header file provide the graphics functions for Borland International's compilers. The \FILTER\M_GRAPHX.C module along with the M_GRAPHX.H header file provide the graphics functions for Microsoft Corporation's compilers. And, finally, the \FILTER\X_GRAPHX.C module and the X_GRAPHX.H header file provide the graphics functions for the Power C compiler from MIX Software, Inc. This is one of the cleanest ways to accomplish the project for various compilers, but there are a few details which must be covered. First, the generic function names used and their compiler-specific counterparts are shown in Table 4.1.

Not all compilers have graphics functions which produce a one-to-one correspondence with the generic functions used in this project. Many times, however, the functions can be carried out in some other manner. For example, the RegisterFont and UnRegisterFont functions used by the Microsoft compiler are incorporated with the initgraph function used to set the VGA mode in the Borland compiler. The MIX Software compiler provides only a standard font of size 8 x 8 pixels; therefore, font registration and variable size are not issues.

Table 4.1 Comparison of compiler-specific graphics functions.

Generic	Borland	Microsoft	MIX Software
ClrScrn	cleardevice	_clearscreen	clrscrn2
GetTextHeight	textheight	_getfontinfo	(none)
GetTextWidth	textwidth	_getfontinfo	(none)
LineTo	lineto	_lineto	line_to
MoveTo	moveto	_moveto	move_to
OutText	outtext	_outgtext	plots,plotch
Rectangle	rectangle	_rectangle	box
RegisterFont	(none)	_registerfonts	(none)
SetColor	setcolor	_setcolor	pen_color
SetFontChar	settextstyle	_setfont	(none)
SetTextMode	closegraph	_setvideomode	setvmode
SetVGA16Color	initgraph	_setvideomode	setvmode
UnRegisterFont	(none)	_unregisterfonts	(none)

The C code for these modules will not be discussed here; instead it is recommended that the reader review carefully the graphics module to be used. If the compiler to be used is not listed above, the reader will need to generate a specific module for that compiler with the same generic function names as listed above. It is hoped that the modules included with this project will serve as a starting point for such an endeavor.

4.4.2 Generating the Filter Design Program Executable

In order to generate the executable file for the filter design program, each of the modules of the project must be successfully compiled and linked together. Each compiler has a different technique to accomplish this aim, but most define either a project file (*.PRJ) or a make file (*.MAK) in order to automate the process. It is recommended that the *large* memory model be used in compiling the project and that a stack size of at least 5 Kbytes be specified for linking.

There are a few details which must be considered in order to accommodate the various compiler-specific graphics functions. First, in the \FILTER\F_SCREEN.H include file, the proper compiler-specific include file must be given in the last line of the #INCLUDES section. (The include file should be B_GRAPHX.H, M_GRAPHX.H, X_GRAPHICS.H or a graphics include file for another compiler which has been written by the user.) Second, the proper compiler-specific module must be included in the project or make file. (The project or make file should include B_GRAPHX.C, M_GRAPHX.C, X_GRAPHX.C, or a graphics module written by the user.) And, finally, the B_GRAPHX.H and M_GRAPHX.H specify that the font file to be used with the program will be in the same directory as the executable. If that is not the case, the path to the font file must be specified.

4.4.3 Final FILTER Program

The final listing of the FILTER main function is too long to present in this text, but it can be found in the \FILTER\FILTER.C module on the software disk. It can be viewed with any text editor or word processor. There are a couple of additions to the code in addition to the combination of the fragments of previous code shown in earlier sections. First, the function has been modified to include a display of the program name and copyright information upon startup and periodically throughout the program.

Second, the program has been modified to run continuously to allow for more than one filter design at a session. After the filter has been designed, the frequency response can also be calculated and viewed in a variety of ways without having to redesign the filter each time. These additions have been implemented by using two `while` loops which will run continuously until the user specifies an end to the design process.

And, finally, a function called `Clean_Up` has been added to free all memory which may have been allocated during the execution of the program and to close any open files. This `Clean_Up` function is used in two different ways. If it is called as a result of an error in the program, it will use the indicator `all` and will clean up all allocated memory and close all open files. However, if it is called when the user wants to run another frequency response on a filter already designed, it will use an indicator `rpsp_only` and will free only the `Resp_Params` and `Scrn_Params` variables. If, during the normal operation of the program, the user wishes to design a new filter, then `Clean_Up` is called with a `fpdb_only` indicator and the `Filt_Params` variables as well as `descript` and `basename` are freed. In addition, all open files are closed. `Clean_Up` can also be found in the \FILTER\FILTER.C module.

4.5 CONCLUSION

At this point in the text, we have put together the complete structure of a filter design program. The digital filter design and analysis functions have yet to be developed, but we have structured our program in a way which will allow them to be added in a clean, efficient manner. We will discuss those functions in the latter half of this text, but before we do, we will discuss in the next chapter the method of implementing our analog filters by using active filters.

Chapter 5

Analog Filter Implementation Using Active Filters

In this chapter, we will discuss the implementation of the analog approximation functions which we developed and verified in previous chapters. Using active filters to implement the transfer functions is a very popular method today because of the natural correspondence between the analog circuit and the mathematical function. We will not discuss the derivation of the transfer functions for active filters because the development of such circuit analysis techniques is beyond the scope of this text. However, a number of suitable references are given in the analog active filter texts of Appendix A for those who are interested in the derivations. Instead, the circuit topology and transfer function for several common active filters will be presented before determining the component values for each circuit. We will develop implementation procedures for each of the filter selectivities discussed in previous chapters as well as discuss some of the important implementation issues. In addition, we will develop C code to automate the process of analog active filter implementation. The program developed will not only determine the necessary component values for the active filter, but will also generate the text input file to be used by PSpice for a complete computer analysis of the final analog circuit.

5.1 Implementation Procedures for Analog Filters

Each of the active filters discussed will be composed of several stages of electronic circuitry consisting of a single operational amplifier (op-amp) and a number of electronic components called resistors (R's) and capacitors (C's). Each of these stages of electronic filtering will have a transfer function which characterizes the relationship of the output voltage to the input voltage as indicated in Equation 5.1.

$$H_c(s) = V_o(s)/V_i(s) \tag{5.1}$$

More specifically, each quadratic factor of the transfer function will have the form as shown in Equation 5.2 where each of the A's and B's in the transfer function will be a function of the R's and C's used in the circuit. (The subscript c is used to indicate that these transfer functions are describing the *circuit* response.)

We also have the approximation function which we have derived to meet a specific set of frequency and attenuation characteristics. The general form of each quadratic factor in the approximation function is shown in Equation 5.3 where the subscript a is used to designate this quadratic factor as an *approximation* function. The a's and b's used as coefficients have already been determined and are numerical constants.

$$H_c(s) = \frac{A_o \cdot s^2 + A_1 \cdot s + A_2}{B_o \cdot s^2 + B_1 \cdot s + B_2} \tag{5.2}$$

$$H_a(s) = \frac{a_o \cdot s^2 + a_1 \cdot s + a_2}{b_o \cdot s^2 + b_1 \cdot s + b_2} \tag{5.3}$$

The implementation process then becomes a matter of equating the two transfer function factors as shown below. Each A and B of the circuit transfer function are equated to the respective a and b of the approximation function and results in several equations which must be solved for appropriate R and C values.

$$H_c(s) = H_a(s) \tag{5.4}$$

We will see this common procedure used throughout the next four sections as we determine the component values needed to implement each of the active filter selectivities.

5.2 Lowpass Active Filters Using Op-amps

There are a number of active filter topologies (circuit configurations) which could be used to implement a lowpass filter. We will limit ourselves to the popular Sallen-Key filter as shown in Figure 5.1 which has a transfer function as described in Equation 5.5. As indicated by the transfer function, this active filter stage can implement one second-order factor of a lowpass filter approximation function. This form is very convenient since it naturally implements the quadratic factor which we have been using for the description of the approximation function. Of course, several of these circuit stages in cascade can be used to implement higher-order functions and, with the addition of a single first-order stage, odd-order filters can be implemented as well.

In comparison to the transfer function of Equation 5.5, Equation 5.7 shows the general form of an all-pole lowpass approximation function such as the Butterworth and Chebyshev functions. (Inverse Chebyshev and elliptic filter approximations will use a different active filter to implement them since they require zeros in the complex plane.) The numerator value a_2 in Equation 5.7 will always be equal to b_2, and G will have a value of unity except for the even-order Chebyshev case.

$$H_{c,L}(s) = \frac{K / R_1 R_2 C_1 C_2}{s^2 + [1/R_1C_1 + 1/R_2C_1 + (1-K)/R_2C_2]\cdot s + 1/R_1R_2C_1C_2} \tag{5.5}$$

where

$$K = 1 + (R_B / R_A) \tag{5.6}$$

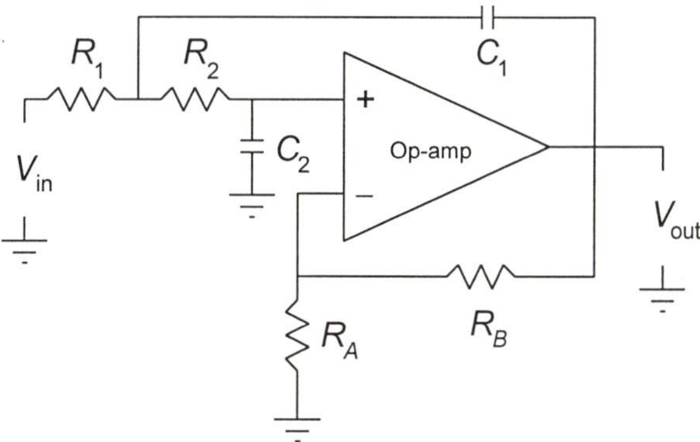

Figure 5.1 Sallen-Key lowpass active filter stage.

$$H_{a,L}(s) = \frac{G\cdot a_2}{s^2 + b_1\cdot s + b_2} \tag{5.7}$$

If we equate Equations 5.5 and 5.7 as indicated in the previous section, we would generate one equation for b_1 and one equation for b_2 (or a_2). However, we have five unknowns (R_1, R_2, C_1, C_2, and K) in our circuit. Since there are more unknowns than equations, we cannot uniquely determine the values of the components required for the active filter. This allows us to select values for three of the components. For example, we could select the values of R_1, R_2, and C_2 if we wanted or the values of C_1, C_2, and K. Another method which we will use is to let $R_1 = R_2$ and $C_1 = C_2$, which effectively eliminates two of the unknowns. In addition we will pick a common value for the capacitors since they have far fewer available values than resistors. This method of picking the R and C values has the advantage of reducing the number of different component values needed to implement the active filter. The disadvantage of this method is that the value of K, which represents the gain for the active filter, will not be unity. However, we will find that this problem can be easily solved.

If we select common values for the primary resistors and capacitors, the transfer function for the active filter then becomes

$$H_{c,L}(s) = K \cdot \frac{1/R^2C^2}{s^2 + [(3-K)/RC]s + 1/R^2C^2} \tag{5.8}$$

By considering the denominators of Equations 5.7 and 5.8, we can determine the two relationships as shown below.

$$b_2 = 1/R^2C^2 \tag{5.9}$$

$$b_1 = (3-K)/RC \tag{5.10}$$

Since we have already picked the value of C, we can solve for the resistor values needed to implement the filter. The results are

$$R = 1/\sqrt{b_2C^2} \tag{5.11}$$

$$K = 3 - (b_1/\sqrt{b_2}) \tag{5.12}$$

and then using Equation 5.6

$$R_B/R_A = 2 - (b_1/\sqrt{b_2}) \tag{5.13}$$

Usually, R_A will be picked as some convenient value and R_B will then be calculated.

The only adjustment remaining is to equate the numerator terms of the two transfer functions. Notice that in this active filter case, there exists a gain of K at $\omega = 0$, while the approximation function has a gain G. In many cases, amplification is a required part of the filtering system, and therefore the differences in gain can be factored into the overall gain requirement. However, in order to be complete, let's determine a method of adjusting the gain of our implementation to the value G as required by the approximation function.

The value of G will always be less than or equal to one, while the value K will always be greater than or equal to one. Thus it is always necessary to reduce the gain of the active filter to match the gain of the approximation function. The amount of this gain adjustment factor is

$$GA = K_{\text{tot}} \,/\, G_{\text{tot}} \tag{5.14}$$

where K_{tot} and G_{tot} represent the products of all of the K's and G's in the total circuit and approximation functions, respectively. This gain adjustment factor can be implemented by a simple voltage divider at the output of the active filter stage. There are two conditions placed on this voltage divider. The inverse of the voltage divider ratio must match the gain adjustment factor, and the equivalent resistance of the voltage divider as seen by the output load must equal the required filter output resistance. Therefore, assuming an output resistance R_{out}, and a voltage divider network made up of R_{x} and R_{y},

$$R_{\text{out}} = R_x \cdot R_y \,/\, (R_x + R_y) \tag{5.15}$$

$$GA = (R_x + R_y) \,/\, R_y \tag{5.16}$$

which means that

$$R_x = GA \cdot R_{\text{out}} \tag{5.17}$$

$$R_y = GA \cdot R_{\text{out}} \,/\, (GA - 1) \tag{5.18}$$

The circuit of Figure 5.2 shows one stage of the new configuration with the voltage divider output. If the input signal level is very high, we might choose to use the voltage divider on the first stage of the filter in order to reduce distortion in the filter.

If an odd-order approximation factor is to be implemented, an odd-order stage as shown in Figure 5.3 can be used as the first stage of the active filter. This RC filter is followed by a buffer amp so that the output impedance

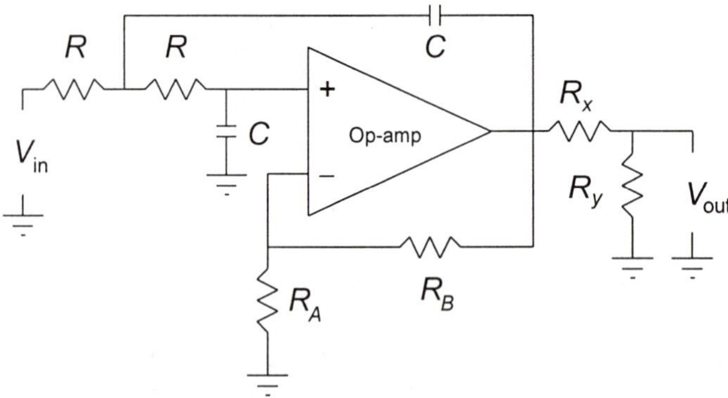

Figure 5.2 Active filter with voltage divider output.

of the RC combination will not affect the input to the next active filter stage. However, if an op-amp is going to be required to implement this first-order factor, we might consider adding a couple more components and implementing a second-order stage instead. In many cases, additional attenuation in the stopband would be welcome, and the additional cost is slight.

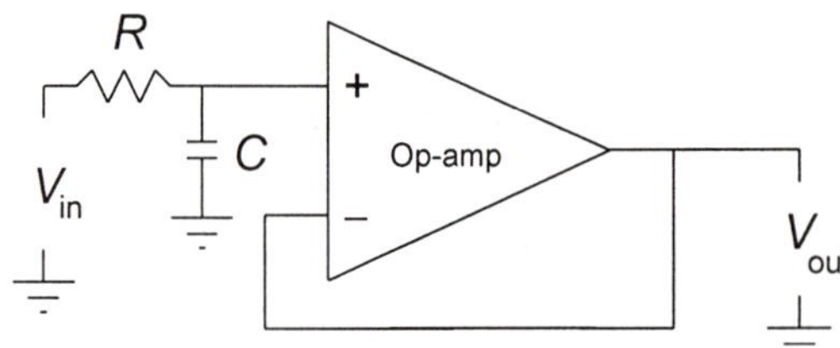

Figure 5.3 First-order lowpass filter with buffer amp.

The transfer function for the first-order stage is given in Equation 5.19. This transfer function must match a first-order approximation function as given in Equation 5.20.

$$H_c(s) = \frac{1/RC}{s + 1/RC} \tag{5.19}$$

$$H_a(s) = \frac{G \cdot a_2}{s + b_2} \tag{5.20}$$

Again, the values of a_2 and b_2 will be identical and G will have a value of one. Then using the value of C that has already been picked,

$$R = 1/b_2 C \tag{5.21}$$

We can use an example to illustrate how to use the information presented in this section to design an active lowpass filter.

Example 5.1 Butterworth Lowpass Active Filter Design

Problem: Determine the resistor and capacitor values to implement a Butterworth lowpass active filter to meet the following specifications:

$a_{pass} = -1$ dB, $a_{stop} = -50$ dB, $f_{pass} = 1000$ Hz, and $f_{stop} = 4000$ Hz

Solution: First, we find that a fifth-order filter is required. The resulting unnormalized approximation function can then be determined to be

$$H_a(s) = \frac{7.1922 \cdot 10^3 (51.728 \cdot 10^6)^2}{(s + 7192.2)(s^2 + 4445.0 \cdot s + 51.728 \cdot 10^6)(s^2 + 11637 \cdot s + 51.728 \cdot 10^6)}$$

The three factors in the denominator can now be matched to three active filter stages. By picking $C = 0.01$ μF and $R_A = 10$ kΩ, the remaining values for each stage can be calculated from Equations 5.11 – 5.13 and 5.21. The first-order stage does not require an R_B value.

m	R_m	K_m	R_{Bm}
0	13.90 kΩ	—	—
1	13.90 kΩ	2.3820	13.82 kΩ
2	13.90 kΩ	1.3820	3.820 kΩ

In order to achieve a gain of exactly 1 at $\omega = 0$, we can use a voltage divider at the output. First, we determine K_{tot}

$$K_{tot} = \prod_m K_m = 3.2919$$

This value along with the fact that G_{tot} has a value of one allows us to determine that $GA = 3.2919$. The following resistor divider values can then be determined assuming a required output impedance of 10 kΩ.

$$R_x = 32.92 \text{ k}\Omega, \quad R_y = 14.36 \text{ k}\Omega$$

The resulting active filter is shown in Figure 5.4. We will generate the C code for the determination of these component values in a later section of this chapter.

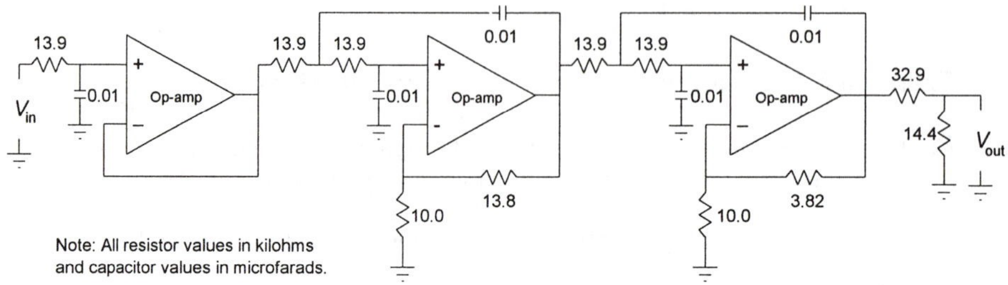

Figure 5.4 Butterworth lowpass active filter for Example 5.1.

5.3 HIGHPASS ACTIVE FILTERS USING OP-AMPS

We will also use a Sallen-Key circuit configuration for implementing a highpass filter. The circuit for implementing a second-order factor is shown in Figure 5.5, with the transfer function for the stage shown in Equation 5.22. We can see that this highpass configuration is the same as the lowpass case except that the primary R's and C's have changed positions, although R_A and R_B remain in their original position.

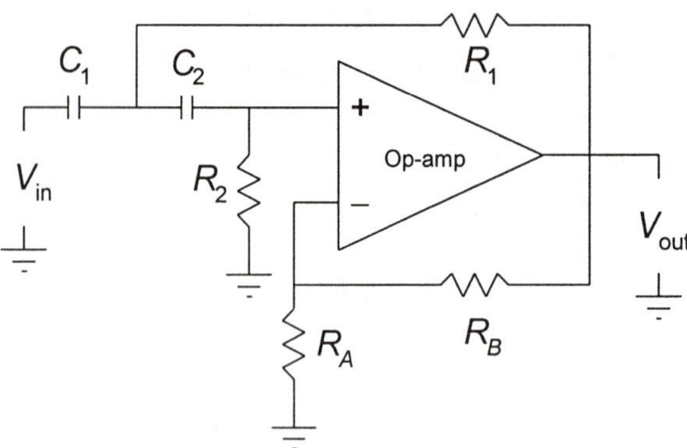

Figure 5.5 Sallen-Key highpass active filter stage.

$$H_{c,H}(s) = \frac{K \cdot s^2}{s^2 + [1/R_2C_2 + 1/R_2C_1 + (1-K)/R_1C_1] \cdot s + 1/R_1R_2C_1C_2} \quad (5.22)$$

where again

$$K = 1 + (R_B / R_A) \quad (5.23)$$

We can use the same procedure in selecting the component values as we used in the lowpass case by letting $R_1 = R_2$ and $C_1 = C_2$. The result is shown in Equation 5.24 where we see that the denominator is identical to the lowpass case in Equation 5.8.

$$H_{c,H}(s) = \frac{K \cdot s^2}{s^2 + [(3-K)/RC] \cdot s + 1/R^2C^2} \quad (5.24)$$

This equation can then be matched to the quadratic factors from the approximation functions which have the form shown in Equation 5.25.

$$H_{a,H}(s) = \frac{G \cdot s^2}{s^2 + b_1 \cdot s + b_2} \quad (5.25)$$

When the equation for the approximation function is compared to the equation for the active filter stage, we see that the relationships between terms are identical to those of the lowpass case. Therefore, if we pick the same value of capacitor, the resistor values will be identical to those of the lowpass case as shown below.

$$R = 1/\sqrt{b_2C^2} \quad (5.26)$$

$$R_B / R_A = 2 - (b_1 / \sqrt{b_2}) \quad (5.27)$$

The adjustment of gain for the highpass case is handled in a similar manner to the lowpass case. As Equation 5.24 indicates, the K value of each stage can be determined by allowing the frequency to approach infinity, as opposed to zero in the lowpass case. K_{tot} can then be easily determined and with the approximation function gain, the gain adjustment factor GA can be determined as shown previously in Equation 5.14. A resistive voltage divider is used at the output of the last stage of the active filter. If necessary, a buffer amplifier can be used after this voltage divider if the impedance of the network being driven by the filter is too small.

If an odd-order highpass approximation factor is to be implemented, an active filter stage as shown in Figure 5.6 can be used as the first stage of the active filter. The only difference between this stage and the lowpass first-order stage is the interchange of R and C values.

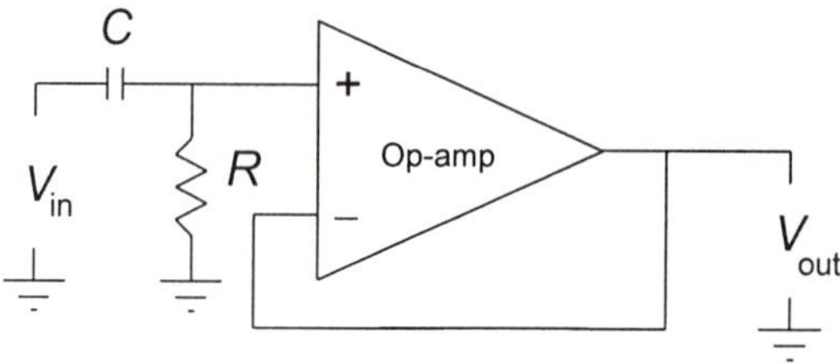

Figure 5.6 First-order highpass filter with buffer amp.

The transfer function for this stage is given in Equation 5.28, while the approximation function for a first-order highpass factor is shown in Equation 5.29.

$$H_c(s) = \frac{s}{s + 1/RC} \tag{5.28}$$

$$H_a(s) = \frac{G \cdot s}{s + b_2} \tag{5.29}$$

As in the second-order case, the resistor value will be the same as the lowpass value assuming that the same value of capacitor is picked.

$$R = 1/b_2 C \tag{5.30}$$

Example 5.2 Chebyshev Highpass Active Filter Design

Problem: Determine the resistor and capacitor values to implement a Chebyshev highpass active filter to meet the following specifications:

$a_{\text{pass}} = -0.5$ dB, $a_{\text{stop}} = -30$ dB, $f_{\text{pass}} = 1000$ Hz, and $f_{\text{stop}} = 400$ Hz

Solution: A fourth-order transfer function as shown below is required. Since this is an even-order Chebyshev, there is an adjustment factor of 0.94406 included in the numerator.

$$H_a(s) = \frac{0.94406 \cdot s^4}{(s^2 + 2071.9 \cdot s + 37.121 \cdot 10^6) \cdot (s^2 + 14926 \cdot s + 110.77 \cdot 10^6)}$$

The two quadratic terms can be matched to two active filter stages by again picking $C = 0.01$ µF and $R_A = 10$ kΩ. The other circuit values can be calculated as shown below.

m	R_m	K_m	R_{Bm}
0	16.41 kΩ	2.6599	16.6 kΩ
1	9.502 kΩ	1.5818	5.82 kΩ

In order to achieve a gain of exactly 1 at $\omega = \infty$, we can use a voltage divider at the output. First, we determine the gain adjustment factor for the filter which in this case includes not only the K_m factors for each quadratic, but also the approximation function's gain of 0.94406.

$$GA = (\prod_m K_m) \,/\, 0.94406 = 4.4567$$

Assuming a equivalent resistance of 10 kΩ, the resistor divider values can be determined with the resulting filter shown in Figure 5.7.

$$R_x = 44.57 \text{ k}\Omega, \quad R_y = 12.89 \text{ k}\Omega$$

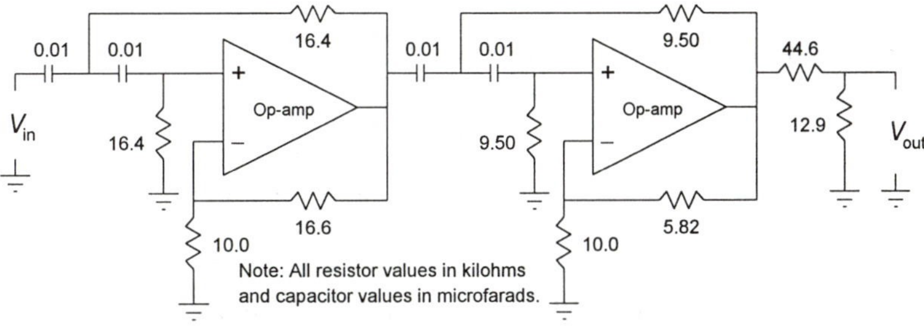

Figure 5.7 Chebyshev highpass active filter for Example 5.2.

5.4 Bandpass Active Filters Using Op-amps

Figure 5.8 shows a Sallen-Key active filter stage which implements a second-order bandpass transfer function. The transfer function for this bandpass stage is given in Equation 5.31.

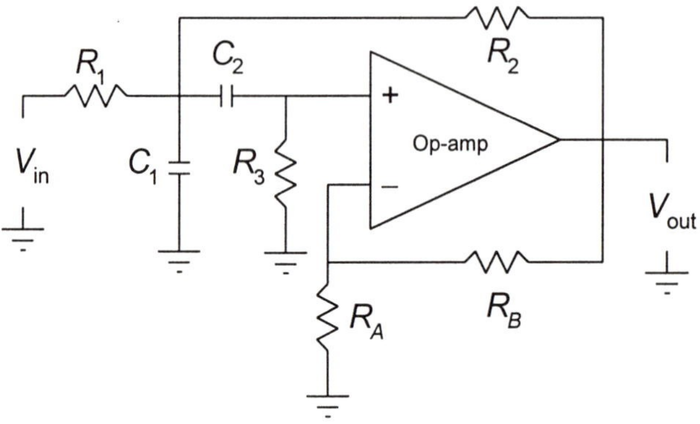

Figure 5.8 Sallen-Key bandpass active filter stage.

$$H_{c,P}(s)=\frac{K\cdot s/R_1C_1}{s^2+\left[\frac{1}{R_1C_1}+\frac{1}{R_3C_1}+\frac{1}{R_3C_2}+\frac{(1-K)}{R_2C_1}\right]\cdot s+\frac{R_1+R_2}{R_1R_2R_3C_1C_2}} \tag{5.31}$$

where again

$$K=1+(R_B/R_A) \tag{5.32}$$

We can simplify this function by letting $R_1 = R_2 = R_3$ and $C_1 = C_2$. This simplified function is given in Equation 5.33.

$$H_{c,L}(s)=\frac{K\cdot s/RC}{s^2+[(4-K)/RC]\cdot s+2/R^2C^2} \tag{5.33}$$

This bandpass transfer function can now be matched to the general form of the approximation factor as shown in Equation 5.34.

$$H_{a,H}(s) = \frac{a_1 \cdot s}{s^2 + b_1 \cdot s + b_2} \tag{5.34}$$

After matching equivalent denominator terms, the following equations emerge,

$$b_1 = (4 - K)/RC \tag{5.35}$$

$$b_2 = 2/R^2C^2 \tag{5.36}$$

which leads to

$$R = \sqrt{2/b_2C^2} \tag{5.37}$$

$$K = 4 - \sqrt{2b_1^2/b_2} \tag{5.38}$$

$$R_B/R_A = 3 - \sqrt{2b_1^2/b_2} \tag{5.39}$$

After these calculations are made, the overall gain of the active filter must be determined. This is a more difficult task than for the lowpass and highpass cases since the gain adjustment must be determined at the center frequency of the passband ω_o. If we refer again to Equations 5.33 and 5.34, we see that they will have identical denominator coefficients since we have just derived the equations to guarantee that. However, the numerators differ by the constants which are present. The gain adjustment factor for each stage is then the ratio of the two numerator constants as shown in Equation 5.40. The total gain adjustment is then the product of these stage gain adjustments.

$$GA = \prod_m K_m / (a_{1m} R_m C_m) \tag{5.40}$$

Once the total gain adjustment has been determined, a voltage divider stage at the output of the active filter can be used for compensation.

Example 5.3 Butterworth Bandpass Active Filter Design

Problem: Determine the resistor and capacitor values to implement a Butterworth bandpass active filter to meet the following specifications:

$a_{pass} = -1.5$ dB, $a_{stop} = -28$ dB, $f_{pass1} = 1000$ Hz,
$f_{pass2} = 2000$ Hz, $f_{stop1} = 500$ Hz, and $f_{stop2} = 4000$ Hz

Solution: A third-order equivalent lowpass filter is needed, which indicates that a sixth-order bandpass function will result as shown below.

$$H_a(s) = \frac{(7282 \cdot s)^3}{(s^2 + 7282 \cdot s + 78.96 \cdot 10^6)(s^2 + 2402s + 38.88 \cdot 10^6)(s^2 + 4880s + 160.3 \cdot 10^6)}$$

By picking C = 0.01 μF and R_A = 10 kΩ, the remaining values can be determined by matching the three quadratic terms.

m	R_m	K_m	R_{Bm}	GA_m
0	15.92 kΩ	2.8410	18.41 kΩ	2.4512
1	22.68 kΩ	3.4550	24.55 kΩ	2.0919
2	11.17 kΩ	3.4550	24.55 kΩ	4.2481

The combined gain produced by the active filter stages at $\omega = \omega_0$ is determined to be 21.783, which can be compensated by a voltage divider at the output of the filter. If the output resistance is to be 10 kΩ, the voltage divider resistor values are

$$R_x = 218 \text{ k}\Omega, \quad R_y = 10.5 \text{ k}\Omega$$

The resulting bandpass filter is shown in Figure 5.9.

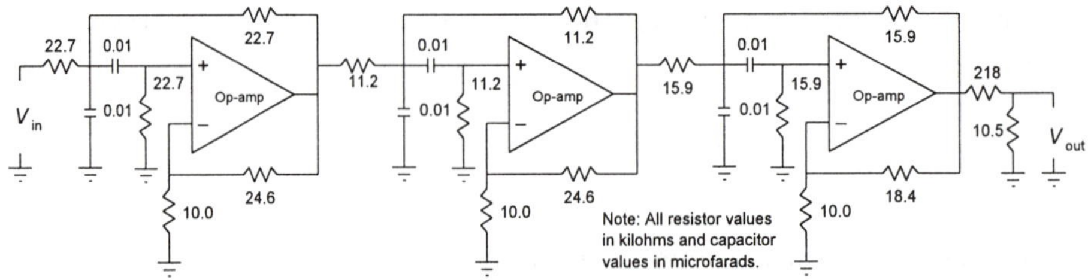

Figure 5.9 Butterworth bandpass active filter for Example 5.3.

5.5 Bandstop Active Filters Using Op-amps

Figure 5.10 shows a twin-tee bandstop active filter stage which can implement a variety of second-order functions. The admittance labeled *Y* can represent a conductance *G*, or a susceptance *sC*, or can have zero value (not be present at all).

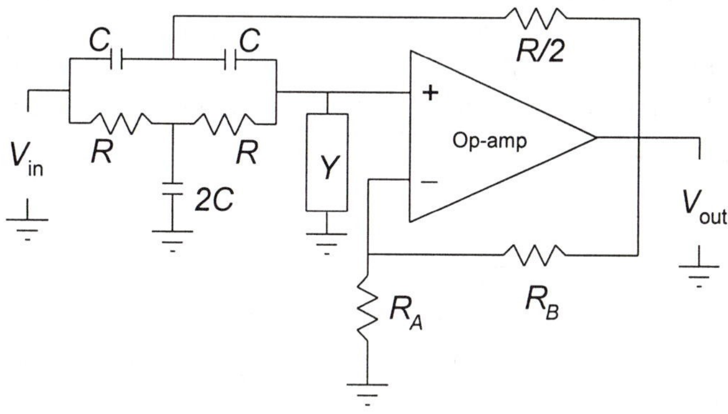

Figure 5.10 Twin-tee bandstop active filter stage.

The general form of the transfer function for this filter is given in Equation 5.41.

$$H_{c,S}(s) = \frac{K \cdot \left[s^2 + \dfrac{1}{R^2C^2} \right]}{s^2 + \left[\dfrac{4 - 2K + 2RY}{RC} \right] \cdot s + \dfrac{1 + 2RY}{R^2C^2}} \tag{5.41}$$

where again

$$K = 1 + (R_B / R_A) \tag{5.42}$$

Equation 5.43 rewrites this equation in terms of the pole frequency ω_p and the zero frequency ω_z.

$$H_{c,S}(s) = \frac{K \cdot (s^2 + \omega_z^2)}{s^2 + (\omega_p / Q) \cdot s + \omega_p^2} \tag{5.43}$$

The transfer function of Equation 5.43 must be matched to the general form of a bandstop approximation function as shown in Equation 5.44.

$$H_{a,S}(s) = \frac{G \cdot (s^2 + a_2)}{s^2 + b_1 \cdot s + b_2} \tag{5.44}$$

Depending on the value of Y selected, ω_z may be greater than, equal to or less than ω_p. This will affect the matching of the respective terms in Equations 5.43 and 5.44. The responses for the transfer function will also change as indicated in Figure 5.11. Let's look at how the transfer function changes for the three cases.

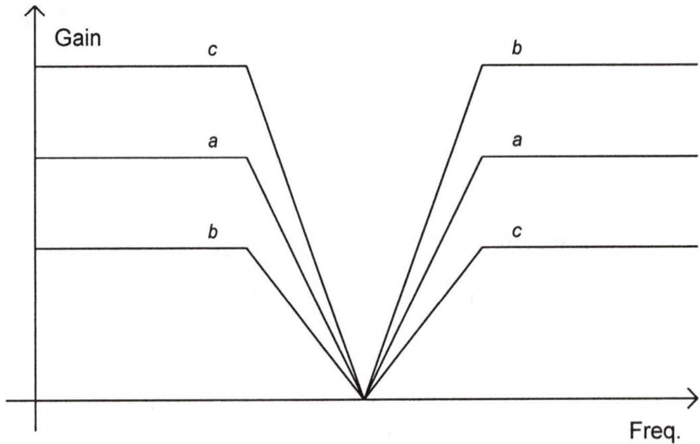

Figure 5.11 Bandstop filter responses.

First, we will consider the case where no Y element is present ($Y = 0$) in which case the transfer function can be simplified as shown in Equation 5.45. Note that this function has the condition that $\omega_z = \omega_p$ ($a_2 = b_2$) which will produce a bandstop response as shown in Figure 5.11(a). Notice that the response has the condition that the upper and lower passbands have equal gain.

$$H_{c,S0}(s) = \frac{K \cdot \left[s^2 + \dfrac{1}{R^2C^2} \right]}{s^2 + \left[\dfrac{4 - 2K}{RC} \right] \cdot s + \dfrac{1}{R^2C^2}} \tag{5.45}$$

The selection of components to implement this type of response is relatively easy as we match the terms of Equation 5.44 and 5.45. By first picking a value of C, the results are

$$R = 1\Big/\sqrt{a_2 C^2} \tag{5.46}$$

$$K = 2 - \sqrt{b_1^2/4b_2} \tag{5.47}$$

$$R_B/R_A = 1 - \sqrt{b_1^2/4b_2} \tag{5.48}$$

Next, in the case where $Y = G_o = 1/R_o$ the resulting transfer function is given in Equation 5.49. Note that this function has $\omega_z < \omega_p$ $(a_2 < b_2)$, which will produce a bandstop response as shown in Figure 5.11(b). This is sometimes referred to as a "highpass notch" filter because the gain of the upper passband is larger than that of the lower passband.

$$H_{c,S1}(s) = \frac{K \cdot \left[s^2 + \dfrac{1}{R^2C^2} \right]}{s^2 + \left[\dfrac{(4 - 2K + 2R/R_o)}{RC} \right] \cdot s + \dfrac{(R_o + 2R)/R_o}{R^2C^2}} \tag{5.49}$$

The selection of components to implement this type of response is again determined by matching terms and by first picking a value of C, the results are

$$R = 1\Big/\sqrt{a_2 C^2} \tag{5.50}$$

$$R_o = 2R/[(b_2/a_2) - 1] \tag{5.51}$$

$$K = 2 + \left(b_2 - a_2 - b_1\sqrt{a_2}\right)\Big/(2a_2) \tag{5.52}$$

$$R_B/R_A = 1 + (b_2 - a_2 - b_1\sqrt{a_2})/(2a_2) \tag{5.53}$$

Finally, in the case where $Y = sC_o$, the resulting transfer function is given in Equation 5.54. Note that this function has $\omega_z > \omega_p$ ($a_2 > b_2$), which will produce a bandstop response as shown in Figure 5.11c. This is sometimes referred to as a "lowpass notch" filter.

$$H_{c,S2}(s) = \frac{\dfrac{K \cdot C}{(C + 2C_o)} \cdot \left[s^2 + \dfrac{1}{R^2C^2}\right]}{s^2 + \left[\dfrac{(4 - 2K) \cdot C + 2C_o}{RC \cdot (C + 2C_o)}\right]s + \dfrac{C}{R^2C^2 \cdot (C + 2C_o)}} \tag{5.54}$$

The selection of components to implement this type of response is again determined by matching terms and by first picking a value of C, the results are

$$R = 1/\sqrt{a_2C^2} \tag{5.55}$$

$$C_o = C \cdot [(a_2 / b_2) - 1]/2 \tag{5.56}$$

$$K = 2 + (a_2 - b_2 - b_1\sqrt{a_2})/(2b_2) \tag{5.57}$$

$$R_B/R_A = 1 + (a_2 - b_2 - b_1\sqrt{a_2})/(2b_2) \tag{5.58}$$

After these calculations are made, the overall gain of the active filter can be determined by evaluating the transfer function at $\omega = 0$ or $\omega = \infty$. If the gain of the circuit is to be determined at $\omega = 0$, then the value of K for that stage is used except for the highpass notch stage. In that case, an additional multiplication factor of $R_o / (R_o + 2R)$ should be included as seen from Equation 5.49. If the gain is to be determined at an infinite frequency, then the lowpass notch stage will have a gain which must be increased by a value of $C / (C + 2C_o)$, as indicated by Equation 5.54. These values will be the same and can be included in the circuit using a voltage divider.

Example 5.4 Chebyshev Bandstop Active Filter Design

Problem: Determine the resistor and capacitor values to implement a Chebyshev bandstop active filter to meet the following specifications:

$a_{pass} = -1$ dB, $a_{stop} = -40$ dB, $f_{pass1} = 666.67$ Hz, $f_{pass2} = 1500$ Hz, $f_{stop1} = 909.09$ Hz, and $f_{stop2} = 1100$ Hz

Solution: A third-order lowpass equivalent function is required which indicates that a sixth-order unnormalized bandstop approximation function will be necessary as shown below.

$$H_a(s) = \frac{(s^2 + 39.48 \cdot 10^6)^3}{(s^2 + 10{,}600 \cdot s + 39.48 \cdot 10^6)(s^2 + 811.0s + 17.87 \cdot 10^6)(s^2 + 1792s + 87.21 \cdot 10^6)}$$

By picking $C = 0.01$ μF and $R_A = 10$ kΩ, the remaining values can be calculated by matching terms. One of the stages will have an R_o value, one will have a C_o value, and the other stage will have no Y element.

m	R_m	K_m	R_{Bm}	Y_m
0	15.92 kΩ	1.1569	1.569 kΩ	—
1	15.92 kΩ	2.4619	14.62 kΩ	$C_o = 6.045$ nF
2	15.92 kΩ	2.4619	14.62 kΩ	$R_o = 26.33$ kΩ

The combined gain produced by the active filter stages at ω = 0 or at infinity is 3.17422, which includes the product of the three K values and either the capacitor or resistor ratio of 0.45269. The voltage divider resistor values can be calculated accordingly. The resulting bandstop filter is shown in Figure 5.12.

$$R_x = 31.74 \text{ k}\Omega, \quad R_y = 14.60 \text{ k}\Omega$$

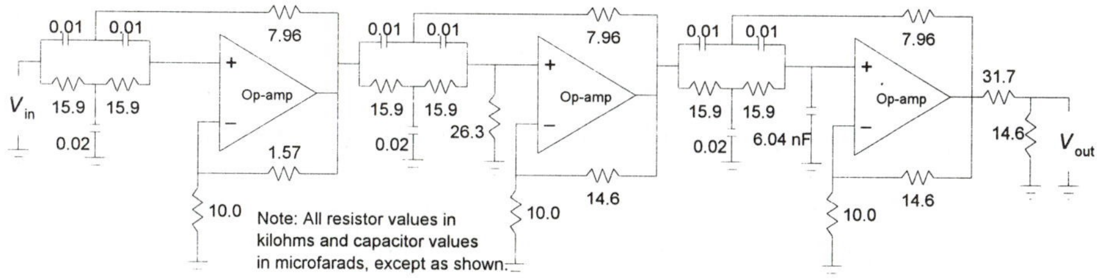

Figure 5.12 Chebyshev bandstop active filter for Example 5.4.

5.6 Implementing Complex Zeros with Active Filters

Up to this point in the chapter, we have not discussed the implementation of rational functions such as the inverse Chebyshev and elliptic approximations. In previous chapters, we discovered that the primary difference between all-pole and rational approximation functions was that the rational functions required complex zeros on the $j\omega$ axis in the s-plane. The general form of a rational function quadratic factor is shown below. If the a_1 coefficient is zero (as it is for zeros on the $j\omega$ axis), the quadratic factor is identical to the form of a bandstop function derived for the all-pole response in the last section.

$$H_{a,S}(s) = \frac{G \cdot (s^2 + a_1 \cdot s + a_2)}{s^2 + b_1 \cdot s + b_2} \tag{5.59}$$

Therefore, we have already developed an active filter form to implement the inverse Chebyshev and elliptic approximations. It is the twin-tee filter presented in the previous section. We can use that form for any of the selectivities (lowpass, highpass, bandpass, or bandstop) when designing inverse Chebyshev or elliptic filters.

However, there are a few special concerns which must be considered when implementing filter approximation functions with complex zeros. If a lowpass or highpass approximation function has an odd-order, the first-order factor should still be implemented by the appropriate first-order stage discussed previously. And in the case of a bandpass approximation, if the approximation function was derived from an odd-order lowpass function, then the quadratic factor associated with the first-order factor should be implemented by a Sallen-Key bandpass active filter section. The other quadratic factors of the bandpass function and the quadratic factors of the lowpass and highpass functions are then implemented by the twin-tee notch filter. Of course, all stages of a bandstop filter of any approximation use the twin-tee configuration. Two examples will now be used to demonstrate the procedure of implementing inverse Chebyshev and elliptic approximations.

Example 5.5 Inverse Chebyshev Lowpass Active Filter Design

Problem: Determine the resistor and capacitor values to implement an inverse Chebyshev lowpass active filter to meet the following specifications:

$$a_{\text{pass}} = -1 \text{ dB},\ a_{\text{stop}} = -60 \text{ dB},\ f_{\text{pass}} = 100 \text{ Hz, and } f_{\text{stop}} = 300 \text{ Hz}$$

Solution: In this case a fifth-order filter will be used. We can determine the required approximation function to be

$$H_a(s) = \frac{10.886 \cdot 10^{-3}(865.77)(s^2 + 3.9282 \cdot 10^6)(s^2 + 10.284 \cdot 10^6)}{(s + 865.77)(s^2 + 449.34 \cdot s + 629.45 \cdot 10^3)(s^2 + 1305.7 \cdot s + 698.64 \cdot 10^3)}$$

The two quadratic terms must be matched to two twin-tee active filter stages while the first-order factor can be implemented by a simple RC stage. By again picking $C = 0.01\ \mu\text{F}$ and $R_A = 10\ \text{k}\Omega$, the remaining values can be calculated. Note that in this lowpass case, both twin-tee sections employ a capacitor as the additional admittance element.

m	R_m	K_m	R_{Bm}	Y_m
0	115.5 kΩ	—	—	—
1	50.46 kΩ	3.9129	29.13 kΩ	$C_o = 26.20$ nF
2	31.18 kΩ	5.8634	48.63 kΩ	$C_o = 68.60$ nF

We determine the gain adjustment factor by finding K_{tot} as shown below.

$$K_{\text{tot}} = \prod_m K_m = 22.943$$

If the desired output resistance is 10 kΩ, the following values result for the voltage divider with the final filter shown in Figure 5.13.

$$R_x = 229.4\ \text{k}\Omega, \quad R_y = 10.46\ \text{k}\Omega$$

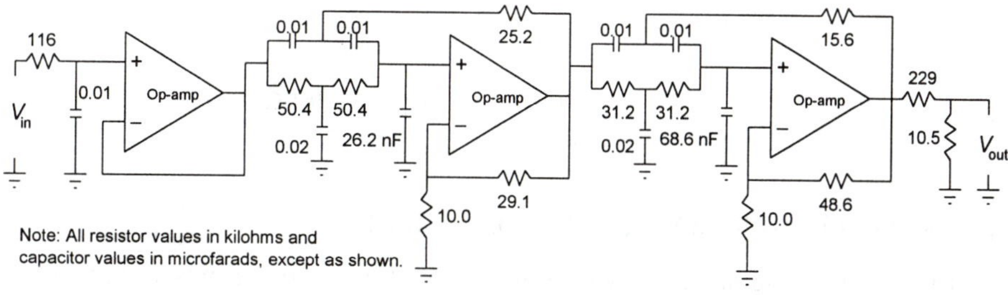

Figure 5.13 Inverse Chebyshev active lowpass filter for Example 5.5.

Example 5.6 Elliptic Bandpass Active Filter Design

Problem: Determine the resistor and capacitor values to implement an elliptic bandpass active filter to meet the following specifications:

$a_{pass} = -1$ dB, $a_{stop} = -60$ dB, $f_{pass1} = 250$ Hz,
$f_{pass2} = 400$ Hz, $f_{stop1} = 100$ Hz, and $f_{stop2} = 1000$ Hz

Solution: The order of the equivalent lowpass filter is 3 which indicates that a sixth-order bandpass approximation function will be necessary. The transfer function is shown below.

$$H_a(s) = \frac{20.82 \cdot 10^{-3} \cdot 469.8 \cdot (s^2 + 50.07 \cdot 10^6) \cdot (s^2 + 311.3 \cdot 10^3)}{(s^2 + 469.8 \cdot s + 3.948 \cdot 10^6)(s^2 + 178.5s + 2.503 \cdot 10^6)(s^2 + 281.6s + 6.227 \cdot 10^6)}$$

These three quadratic terms must be matched to three active filter stages. The stage related to the first-order factor in the *normalized* LP filter is implemented by the standard Sallen-Key bandpass filter. The other two stages are implemented using the twin-tee filters. By picking $C = 0.01$ μF and $R_A = 10$ kΩ, the remaining values can be calculated as shown below. One of the stages will have $R_o = 18.86$ kΩ, one will have $C_o = 95.04$ nF, and the other stage will have no Y element at all.

m	R_m	K_m	R_{Bm}	Y_m
0	71.18 kΩ	3.6656	26.66 kΩ	—
1	14.13 kΩ	11.251	102.5 kΩ	C_o=95.04 nF
2	179.2 kΩ	11.251	102.5 kΩ	R_o=18.86 kΩ

The combined gain produced by the active filter stages at ω = 0 is 4.4406, which can be compensated by a voltage divider after the last stage. The voltage divider resistor values are shown below and are calculated to provide an output impedance of 10 kΩ. The resulting bandstop filter is shown in Figure 5.14.

$$R_x = 44.4 \text{ k}\Omega, \quad R_y = 12.9 \text{ k}\Omega$$

5.7 Analog Filter Implementation Issues

In the previous sections we determined component values necessary to implement our analog filter designs in the form of active filters. These active

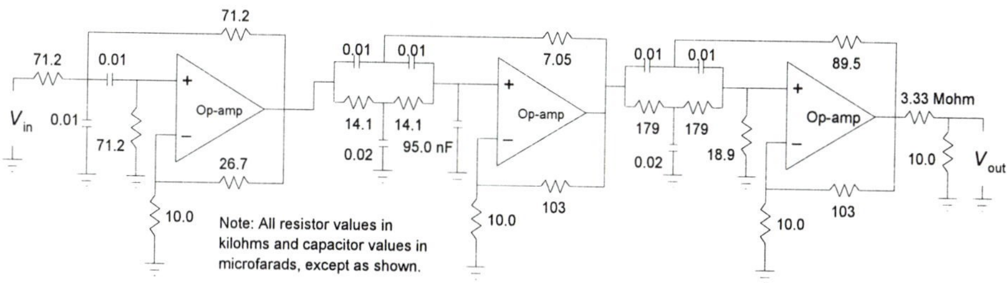

Figure 5.14 Elliptic bandpass active filter for Example 5.6.

filters make use of electronic components which can be manufactured only to a finite accuracy. The components are not perfect when they are first manufactured and can change value because of aging or exposure to heat, chemicals, or humidity. Although we can buy precision components which resist changes, the cost of such components can be expensive. It is most cost effective to specify precision components only for those positions which actually require it. We can determine which components are in critical positions by performing a sensitivity analysis. A sensitivity analysis is the process of finding out how any or all of the characteristics of a filter are affected by each and every component which makes up the filter.

For example, the sensitivity of a function F with respect to x is defined as

$$S_x^F = \frac{x}{F} \cdot \frac{\partial F}{\partial x} = \frac{\partial[\ln(F)]}{\partial[\ln(x)]} \tag{5.60}$$

Note that sensitivity not only considers the change in F as a function of x, but also the nominal values of F and x. Therefore, the sensitivity considers the per unit change of the function with respect to the per unit change of the parameter. Using the definition of Equation 5.60, other common sensitivity relationships can be determined as shown in Equations 5.61 – 5.66. (The value c is considered a constant and G is another function of x.)

$$S_x^{cx} = 1 \tag{5.61}$$

$$S_x^{1/F} = -S_x^F \tag{5.62}$$

$$S_x^{FG} = S_x^F + S_x^G \tag{5.63}$$

$$S_x^{F/G} = S_x^F - S_x^G \tag{5.64}$$

$$S_x^{F^c} = c \cdot S_x^F \tag{5.65}$$

$$S_{x^c}^F = \frac{1}{c} \cdot S_x^F \tag{5.66}$$

In each of the preceding sections, a transfer function for an active filter was given in terms of the resistor and capacitor values used to make up the circuit. In general, those functions can be specified in terms of the pole and zero frequencies and quality factors (Q's) as shown in Equation 5.67.

$$H(s) = K \cdot \frac{s^2 + (\omega_z / Q_z)s + \omega_z^2}{s^2 + (\omega_p / Q_p)s + \omega_p^2} \tag{5.67}$$

If we consider the lowpass active filter of Section 5.2 which has a transfer function as indicated in Equation 5.68, we can easily match terms to identify the pole frequency and Q in terms of the component values as shown in Equations 5.69 and 5.70.

$$H_{c,L}(s) = \frac{K / R_1 R_2 C_1 C_2}{s^2 + [1/R_1C_1 + 1/R_2C_1 + (1-K)/R_2C_2] \cdot s + 1/R_1R_2C_1C_2} \tag{5.68}$$

$$\omega_p = 1/\sqrt{R_1R_2C_1C_2} \tag{5.69}$$

$$Q_p = \frac{1/\sqrt{R_1R_2C_1C_2}}{\dfrac{1}{R_1C_1} + \dfrac{1}{R_2C_1} + \dfrac{1-K}{R_2C_2}} \tag{5.70}$$

The sensitivity of either one of the parameters listed above to any one of the resistor or capacitor values can now be determined. For example,

$$S_{R_1}^{\omega_p} = -0.5 \tag{5.71}$$

which indicates that for every 1.0% *increase* in R_1 there is a 0.5% *decrease* in ω_p. Likewise we can find that

$$S_{R_1}^{Q_p} = -0.5 + Q_p\sqrt{\frac{R_2 C_2}{R_1 C_1}} \tag{5.72}$$

which for equal resistor and capacitor values becomes

$$S_{R_1}^{Q_p} = -0.5 + Q_p \tag{5.73}$$

The selection of R and C values for the active filters plays an important part in the overall sensitivity of the circuit. We do not have space for a full treatment of sensitivity analysis in this text. However, references in Appendix A provide further details on sensitivity and on methods of selecting component values which yield low sensitivities. The text by Daryanani provides a particularly good presentation. In fact, that text presents a different choice for the component values for a Sallen-Key active filter as listed below. By making those selections, it can be shown by substitution into Equation 5.70 that Q_p is no longer a function of R at all and therefore would have a sensitivity of zero with respect to changes in resistance value.

$$K = 1, \quad \mathrm{R}_1 = R_2 = R$$

$$C_1 = \frac{2 \cdot Q_p}{R \cdot \omega_p}, \quad \mathrm{C}_2 = \frac{1}{2 \cdot R \cdot Q_p \cdot \omega_p}$$

A sensitivity analysis provides us with valuable information about the component values used in the circuit. With that information, we can determine which components are critical in controlling variation in key filter parameters. However, it is also very important to perform a worst case analysis on the circuit. This analysis of the circuit would use the sensitivities which have already been determined to set each component to the extreme limit of its tolerance in order to see the effect on the overall circuit performance. For example, all components with negative sensitivities would be set to their lower tolerance limit while all components with positive sensitivities would be set to their upper limits. The circuit would then be analyzed to see if it still met the specifications. Next, all component values would be set to the opposite limit and the test would be repeated. If a circuit passed this test, it should perform up to specifications with the randomly chosen values used when it is assembled. An example of a worst case analysis using PSpice is given in Section 5.8.4.

Another type of statistical test which is less rigorous than a worst case test, but probably more typical of what will happen in real life, is the Monte Carlo simulation. In this test, component values are selected randomly within their tolerance range and the circuit is tested to see if it meets the required specifications. In most software packages, the values can be characterized as having a Gaussian or uniform distribution. A number of Monte Carlo tests are usually run to simulate the variation of component values which will be seen in the normal assembly process.

Another concern for active filter designers is the selection of the active device, the op-amp. It is crucial that the key characteristics of the op-amp match the filter design project. In most present day op-amps, input and output characteristics can be considered nearly ideal, but the frequency response characteristics must be carefully checked. Every op-amp has an upper frequency limit normally specified by its gain-bandwidth product (GBP). This number provides a reference which can be used to determine the open loop gain of the op-amp at any frequency (or vice versa). For example, the GBP of a 741 op-amp is 10^6, which indicates that this amplifier has an open loop gain of 1000 at a frequency of 1000 Hz. Or we could use this GBP to predict that an upper frequency limit of 20,000 Hz was available with an open loop gain of 50. The important point to remember, however, is that the transfer functions of the active filters where derived assuming that the gains of the op-amps used were very large for all frequencies of interest in the design. Therefore, whether or not the active filter is a lowpass, highpass, bandpass, or bandstop filter, the open loop gain of the op-amp must be large (greater than 1000) for all frequencies of interest. These frequencies of interest include not only the passbands, but also the stopbands.

5.8 C Code for Active Filter Implementation

We are now at a point in this chapter where we are ready to develop the C code to aid in the implementation of these analog active filters. We will develop the code necessary for the calculation of component values and the generation of PSpice text files. These text files will serve as input to PSpice which will analyze the active filters. Before we start generating code, however, we need to develop an outline of the project, which will enable us to plan our project more effectively. An outline of our project is listed below indicating the major components.

Analog Active Filter Implementation Project

I. Determine filter parameters.
II. Calculate component values based on parameters.
III. Display component values on screen.
IV. Write PSpice text file.

Although this outline provides us with a large scale view of the project, we will need further details in order to break the project into small enough pieces for individual functions to be written in C. As shown below, by adding an additional level of description to our outline, we have a more defined project.

Analog Active Filter Implementation Project

I. Determine filter parameters.
  A. Determine which file contains filter parameters.
  B. Read the parameters from the file.
  C. Display the filter parameters for user verification.
II. Calculate component values based on parameters.
  A. Calculate components for lowpass, if appropriate.
  B. Calculate components for highpass, if appropriate.
  C. Calculate components for bandpass, if appropriate.
  D. Calculate components for bandstop, if appropriate.
III. Display component values on screen.
IV. Write PSpice text file.
  A. Write text file for lowpass, if appropriate.
  B. Write text file for highpass, if appropriate.
  C. Write text file for bandpass, if appropriate.
  D. Write text file for bandstop, if appropriate.

As indicated in this project outline, our first objective is to determine the file which holds the parameters of the filter to be implemented. We then read the parameters from the file and display them so that the user can verify that the correct file is being processed. We then proceed to calculate the component values required for the specific filter required. After the component values have been displayed, the PSpice text file can be written for later evaluation.

All of the functions required to implement this project are contained in the \ANALOG\ANALOG.C module with the associated include file ANALOG.H. The ANALOG `main` function for our project is shown in Listing 5.1. (Because of its length, some routine code has be omitted.) After an opening screen is displayed describing the program, the screen is cleared and a header at the top of the screen is presented. At that point, we use `Get_ANALOG_Basename` to retrieve the name of the data file holding the filter parameters for the filter to be implemented.

```
/*==========================================================
  MAIN()
==========================================================*/
void main(void)
{ /* Declare and initialize variables */
  { code omitted }
  /*  Display the opening screen */
  { code omitted }
  while(1)
  { /*  Get filenames from user and
        alloc memory for the structures */
    { code omitted }
    /*  Read and display filter parameters */
    Error = Read_Filter_Params(FP,descript,filename);
    if(Error)
    { Clean_Up(filename,FP,RC);
      printf("\n\tFile not found or incorrect type!");
      Bail_Out("Error in Read_Filter_Params!",Error);
    }
    Display_Filt_Params(FP,descript);
    Pause();
    /*  Get user specified parameters */
    { code omitted }
    /*  Calculate and display component values */
    Error = Calc_Components(FP,RC,C,Ra);
    if(Error)
    { Clean_Up(filename,FP,RC);
      Bail_Out("Error in Calc_Components!",Error);
    }
    Display_Components(RC,descript);
    /*  Write data file for computer analysis */
    Error = Write_Circ_File(FP,RC,filename,descript,
                                       fstart,fstop);
```

```
        if(Error)
        { Clean_Up(filename,FP,RC);
          Bail_Out("Error in Write_Circ_File!",Error);
        }
        Clean_Up(filename,FP,RC);
        /*  Run program again? */
        ans = Get_YN(
         "\n\n\tImplement another active filter? (Y/N): ");
        if(ans == 'N')
        { break;}
    }
    Clear_Screen();
    Cursor_Home();
    Bail_Out("",0);
}
```

Listing 5.1 Analog main function.

After the filename of the file has been determined, memory is allocated for two structures which are used to pass information to the various functions in this project. The first structure is the familiar Filt_Params structure, which was used in the FILTER program. The second is a new structure used to hold the component values for the active filters to be designed. This RC_Comps structure is shown below and includes the voltage divider resistors R_x and R_y as well as the number of stages in the filter. The other variables contained in the structure are pointers to arrays because these variables will be present in each stage of the filter and we will not know at the time of the program execution how many stages will be present. The components included are the primary R and C values, the gain control resistors R_A and R_B, and the bandstop admittance parameters R_o and C_o.

```
typedef struct
{ double  Rx,Ry,      /*  Voltage divider values */
          *R,*C,      /*  Primary R-C values */
          *Ra,*Rb,    /*  Feedback resistors */
          *Ro,*Co;    /*  Twin-tee addl components */
  int     stages;     /*  Number of stages */
} RC_Comps;
```

After the successful allocation of memory for the structures, we proceed to read in the filter parameters specified in the input file using `Read_Filter_Params`. If any error occurs in reading the file, we will exit the program. Otherwise, we will display the filter's parameters as read from the file using `Display_Filt_Params` which displays only the filter parameters, and not the coefficients. (The function has been included in the ANALOG.C module for convenience.) This display is both a reminder to the user of the filter specifications and a verification that the correct file is being processed.

At this point in the `main` function, we again clear the screen and ask for data from the user. In particular, we need the common value of capacitor and R_A resistor to use in each of the filter's stages. In addition, the starting frequency and the number of decades are solicited and used to calculate the ending frequency for the PSpice analysis. We use the `Get_Double` function, which was developed earlier in the text, to request this information from the user.

The component values can then be calculated and displayed using the `Calc_Components` and `Display_Components` functions. After the user has finished viewing the component values, the PSpice data file is written using the `Write_Circ_File` function. The user is then asked if another active filter should be implemented. Other formats for this data file can easily be accommodated if the file structure used in this program is not appropriate. If any errors occur in the calculation of the components or the writing of the PSpice file, `Bail_Out` will be used to exit the program with the appropriate message.

5.8.1 Determining and Displaying Filter Parameters

We can begin our project by implementing item I.A of our project outline. The `Get_ANALOG_Basename` function is used to obtain the filename information. The function displays a short message indicating that the user is to enter the name of the binary data file which was written during the design of the filter. However, the .DTB extension is not to be included. This allows for the easy addition of the .CIR extension used for the PSpice data file later.

Item I.B of the project outline is implemented by the `Read_Filter_Params` function in the \ANALOG\ANALOG.C module. This function is effectively the inverse of the `Save_Filter_Params` function which is part of the \FILTER\F_DESIGN.C module. In this function, the binary data from the filter parameter file is read in and transferred to the `Filt_Params` structure. The basename of the file is augmented by adding the extension of .DTB, and the file is opened for binary reading. All of the parameters of the filter are read and then stored in the `Filt_Params` structure.

The display of the filter parameters as required in item I.C in our project outline is handled by the `Display_Filt_Params` function, which was developed earlier in this text. The function definition is included in the \ANALOG\ANALOG.C module so that the very large F_DESIGN.C module (which requires a number of other modules for full definition) is not required in our project.

5.8.2 Calculation and Display of Component Values

Once we have the filter parameters for the filter to be designed, we can calculate the components necessary to implement the active filter as indicated in items II.A–D in our project outline. We begin to implement these items by calling a function called `Calc_Components`. This function (which is contained in the \ANALOG\ANALOG.C module and can be viewed using a text editor or word processor) will determine which of several other functions will be the proper function to call based on the filter's selectivity. If a lowpass filter is being implemented, then the `Calc_LP_Comps` function will be called, while if a bandpass filter is being implemented, the `Calc_BP_Comps` function will be called. However, no matter which function is to be called, memory must be allocated for the arrays of component values in the structure. This memory allocation can be done prior to calling the individual functions and thereby eliminate the requirement of memory allocation in each of the individual functions.

The component calculation functions are very similar to one another although the equations for component values are somewhat different. For that reason, only the `Calc_LP_Comps` function shown in Listing 5.2 will be discussed here. All of the component calculation functions are contained in the \ANALOG\ANALOG.C module included on disk with this text. The work within the function begins by initializing the variable `K_Total`, which will store the product of all stage K's. Then, if a first-order stage is required, the R and C values for it are calculated and `start` is set to 1. Then, the component values for the stages implementing the quadratic factors are calculated within the `for` loop, which has a starting index of `start`. The coefficients from the `Filt_Params` structure are retrieved, and the common values of C and R_A are placed in the `RC_Comps` structure. Then a determination has to be made within a switch statement as to whether a filter with complex conjugate zeros will be required. If the zeros are not required, the calculations are made and the process continues. However, if complex conjugate zeros are required, it must be decided whether a resistor, capacitor or no additional value is necessary.

```
/*=====================================================
  Calc_LP_Comps() - calculates the component values
  for lowpass analog active filter.
  Prototype:  int Calc_LP_Comps(Filt_Params *FP,
         RC_Comps *RC,double C,double Ra);
  Return:     error value
  Arguments:  FP - ptr to struct holding filter params
              RC - ptr to struct holding RC components
              C -  capacitor value for all stages
              Ra - feedback resistor for all stages
=====================================================*/
int Calc_LP_Comps(Filt_Params *FP,RC_Comps *RC,
                                  double C,double Ra)
{ int     i,start;     /* Loop counter and start pt*/
  double  K,K_Total,   /* K value and total */
          Gain_Adj,    /* Gain adjustment factor */
          a2,a2r,      /* Numerator constants */
          b1,b2,b2r;   /* Denominator constants */

  /*  Initialize K total */
  K_Total = 1;
  /*  If order is odd, determine R & C values for
      first-order stage and set start to 1 */
  start = 0;
  if(FP->order % 2)
  { RC->C[0] = C;
    RC->R[0] = 1 / (C * FP->bcoefs[2]);
    RC->Ra[0] = Ra;
    start = 1;
  }
  /*  Determine values for second-order stages */
  for(i = start; i < RC->stages ;i++)
  { /*  Determine coefficients and roots */
    a2 = FP->acoefs[i*3 + 2];
    a2r = sqrt(a2);
    b1 = FP->bcoefs[i*3 + 1];
    b2 = FP->bcoefs[i*3 + 2];
    b2r = sqrt(b2);
    /*  Set standard values in structure */
    RC->C[i] = C;
```

```
        RC->Ra[i] = Ra;
        /*  Calculate values based on approx type
            B,C use Sallen-Key, E,I use Twin-Tee */
        switch(FP->approx)
        { case 'B':
          case 'C':
            RC->R[i] = 1 / (C * b2r);
            K = 3 - (b1/b2r);
            break;
          case 'E':
          case 'I':
            RC->R[i] = 1 / (C * a2r);
            /*  Find K, Ro, Co dependent on a2,b2 */
            if(a2 > b2)
            {       K = 2 + ((a2-b2-b1*a2r)/(2*b2));
              RC->Co[i] = (((a2/b2) - 1) * C) / 2;
            }
            else if(a2 < b2)
            {       K = 2 + ((b2-a2-b1*a2r)/(2*a2));
              RC->Ro[i] = 2 * RC->R[i] / ((b2/a2) - 1);
            }
            else
            { K = 2 - (b1/(2*a2r));}
            break;
          default:
            return ERR_FILTER;
        }
        /*  If Co = 0, K_Tot only increases by K */
        K_Total *= ( (K * C) / (C + 2*RC->Co[i]) );
        RC->Rb[i] = Ra * (K - 1);
    }
    /*  Make final adjustment of gain and
        calculate voltage divider values */
    Gain_Adj = K_Total / FP->gain;
    if(Gain_Adj == 1.000)
    { return ERR_NONE;}
    RC->Rx = Gain_Adj * R_OUT;
    RC->Ry = Gain_Adj * R_OUT / (Gain_Adj - 1);
    return ERR_NONE;
}
```

Listing 5.2 Calc_LP_Comps function.

The loop will continue its iterations until all stage components have been determined. Once the loop finishes, the gain adjustment factor can be determined and the voltage divider components can be calculated. If the gain adjustment is exactly 1, no values are calculated since a division by zero would result.

The display of the component values is handled by the `Display_Components` function shown in Listing 5.3 and corresponds to item III in our project outline. In this function, we first print a header for the component information, which includes the stage number, the primary resistance value R, and the gain resistance value R_B. A `for` loop is then used to print out the R and R_B values as well as special information. This special information could include the R_o or C_o value for a stage implemented by a twin-tee notch filter or an indication that the stage implements a first-order factor. The presence of a first-order stage is indicated by an R_B value of zero. After all of the stage values have been displayed, the common capacitance value C and gain resistance R_A are shown along with the voltage divider resistance values R_x and R_y. The `Pause` function is then used to allow the user time to view or print the displayed information.

```
/*=====================================================
  Display_Components() - displays the component values
  for analog active filter.
  Prototype:  void Display_Components(RC_Comps *RC,
                                      char *descript);
  Return:     error value
  Arguments:  RC - ptr to struct holding RC components
              descript - pointer to filter description
=====================================================*/
void Display_Components(RC_Comps *RC,char *descript)
{ int i;
  /*  Print description and table column headers */
  printf("\n        %s",descript);
  printf("\n\n        Stage    R Value        ");
  printf("R_B Value          Special");
  printf("\n          =====  ===========    ");
  printf("==========     =================");
  /*  Loop through component values for each stage,
      assume even number of stages, use Rb == 0 to
      indicate odd order and so indicate */
  for(i = 0; i < RC->stages ;i++)
  { printf("\n          %2i     %8.4e    %8.4e",
                                i,RC->R[i],RC->Rb[i]);
```

```
      if(RC->Ro[i] != 0.00)
      {        printf("    Ro = %8.4e",RC->Ro[i]);}
      if(RC->Co[i] != 0.00)
      {        printf("    Co = %8.4e",RC->Co[i]);}
      if(RC->Rb[i] == 0.00)
      {        printf("    1st Order section");}
    }
    /*  Show common values of C and Ra outside of loop
        as well as voltage divider values Rx and Ry */
    printf("\n\n                 Common  ");
    printf("Component Values\n          ");
    printf("====================================");
    printf("\n          C      %8.4e",RC->C[0]);
    printf("    R_A  %8.4e",RC->Ra[0]);
    printf("\n          R_x   %8.4e",RC->Rx);
    printf("    R_y  %8.4e",RC->Ry);
    /*  Pause to view or print */
    printf("\n\n            Press any key to continue");
    Pause();
}
```

Listing 5.3 Display_Components function.

5.8.3 Writing the PSpice Analysis File

After an analog active filter has been designed and the component values have been calculated, the next logical step is to test the circuit. Testing usually includes both computer analysis, where a circuit simulation is performed, and laboratory analysis, where the circuit is built from components and tested with electronic equipment. We can help in the computer evaluation of the filter circuit by preparing the analysis data file necessary for PSpice or some other evaluation tool. This preparation corresponds to items IV.A–D in our project outline.

The `Write_Circ_File` function as shown in Listing 5.4 is used to coordinate the generation of the circuit analysis text file. The pointers to the `Filt_Params` and `RC_Comps` structures are included as arguments for the function as well as the filename and filter description. The first activity in the function is to append an extension of .CIR to the filename and open that file as a text file for writing. All subsequent file output will be sent to this file using the `fprintf` statement and the file pointer `CF`.

The first line of text output is the filter description, which will serve as a title for all analysis output. Then a comment line and input source line are written where an AC voltage source of one volt is specified. Next, we enter a `for` loop which will iterate through the number of stages and write the appropriate circuit section based on the selectivity of the filter. After all of the filter sections have been written, the final voltage divider section is appended, and an appropriate model for the op-amp circuit is specified. (In this version of the function, a subcircuit dependent source model is used to provide a near ideal model of the op-amp.) And finally, the analysis modes are specified using the starting and ending frequencies provide by the user.

```
/*=====================================================
  Write_Circ_File() - writes a text data file which
  can be used as input to a circuit analysis package.
  Prototype:  int Write_Circ_File(Filt_Params *FP,
          RC_Comps *RC,char *basename,char *descript);
  Return:     error value
  Arguments:  FP - ptr to struct holding filter params
              RC - ptr to struct holding RC components
              basename - basename for circ file
              descript - pointer to filter description
=====================================================*/
int Write_Circ_File(Filt_Params *FP,RC_Comps *RC,
                        char *basename,char *descript,
                           double fstart,double fstop)
{ char  filename[MAX_NAME_LEN];    /* text filename */
  int   i,s;        /* loop counter and stage indic */
  FILE  *CF;        /* pointer to text file */

  /*  Set up the filename and open file */
  strcpy(filename,basename);
  strcat(filename,".CIR");
  CF = fopen(filename,"wt");
  if(!CF)  { return ERR_FILE;}
  /*  Write title and source voltage */
  fprintf(CF,"%s",descript);
  fprintf(CF,"\n*    Specify input source:");
  fprintf(CF,"\nVs \t11 \t0 \tAC \t1.0 \t0.0");
  /*  Select sections to write based on selectivity */
  for(i = 0; i < RC->stages ;i++)
  { switch(FP->select)
    { case 'L':
```

```
          Write_LP_Section(i,RC,CF);  break;
        case 'H':
          Write_HP_Section(i,RC,CF);  break;
        case 'P':
          Write_BP_Section(i,RC,CF);  break;
        case 'S':
          Write_BS_Section(i,RC,CF);  break;
        default:
          return ERR_FILTER;
      }
    }
    /*      Add voltage divider section */
    s = (RC->stages + 1) * 10;
    fprintf(CF,"\n*    Voltage divider section");
    fprintf(CF,"\nRx \t%d \t%d \t%8.3E",s+1,s+2,RC->Rx);
    fprintf(CF,"\nRy \t%d \t%d \t%8.3E",s+2,0,RC->Ry);
    /*      Write op-amp description */
    fprintf(CF,"\n*    Sub-circuit model for op-amp");
    fprintf(CF,"\n.SUBCKT \tOPAMP \t1 \t2 \t6");
    fprintf(CF,"\nRin \t1 \t2 \t1E+008");
    fprintf(CF,"\nE1 \t3 \t0 \t1 \t2 \t1E+003");
    fprintf(CF,"\nRx \t3 \t4 \t1E+003");
    fprintf(CF,"\nCx \t4 \t0 \t1E-009");
    fprintf(CF,"\nE2 \t5 \t0 \t4 \t0 \t1E+003");
    fprintf(CF,"\nRout \t5 \t6 \t1E000");
    fprintf(CF,"\n.ENDS");
    /*Write analysis modes */
    fprintf(CF,"\n*    Analysis modes");
    fprintf(CF,"\n.AC \tDEC \t100 \t%8.3E \t%8.3E",
                                                fstart,fstop);
    fprintf(CF,"\n.PROBE");
    fprintf(CF,"\n.END");
    /*      Close the file */
    fclose(CF);
    return ERR_NONE;
}
```

Listing 5.4 Write_Circ_File function.

As an example of one of the functions which generates circuit analysis data files, Listing 5.5 shows `Write_LP_Section`. This function begins by simplifying the form of the components for each stage and then writes a section based on the order of the filter stage and the implementation of the filter stage. If the stage is implementing a first-order factor as indicated by R_B having a value of zero, the necessary component values are written to the file. If the stage is implementing a second-order factor, the stage configuration could be either a Sallen-Key or a twin-tee notch form. If either C_o or R_o is nonzero, the twin-tee notch form is indicated, and the `Write_BS_Section` function is called since it implements the twin-tee notch filter form. Otherwise, the components for a standard Sallen-Key stage are written to the text file.

```
/*=========================================================
  Write_LP_Section() - writes a lowpass filter
  section to the circuit data file.
  Prototype:  void Write_LP_Section(int stage,
                                RC_Comps *RC,FILE *CF);
  Return:     none
  Arguments:  stage - section number of filter
              RC - ptr to struct holding RC components
              CF - ptr to output file
=========================================================*/
void Write_LP_Section(int stage,RC_Comps *RC,
                                          FILE *CF)
{ int    s,t;         /* Stage related variables */
  double R,C,Ra,Rb;   /* Component values */

  /*  Simplify some variables */
  t = stage + 1;                          s = 10 * t;
  R = RC->R[stage];    C = RC->C[stage];
  Ra = RC->Ra[stage]; Rb = RC->Rb[stage];
  /*  Rb == 0 if first-order stage, otherwise
      generate circuit text for second-order */
  if(Rb == 0)
  { fprintf(CF,"\n*    Stage Number %d",stage+1);
    fprintf(CF,"\nR%d\t%d\t%d\t%8.3E",s+1,s+1,s+2,R);
    fprintf(CF,"\nC%d\t%d\t%d\t%8.3E",s+1,s+2,0,C);
    fprintf(CF,"\nRb%d\t%d\t%d\t1",t,s+3,s+11);
    fprintf(CF,
        "\nX%d\t%d\t%d\t%d\tOPAMP",t,s+2,s+3,s+11);
  }
  /*  Generate circuit text for second-order stage
      If Ro and Co != 0, then use BS stage to
      generate elliptic or inv Chebyshev approx,
      otherwise use standard BP configuration */
  else
```

```
      {if( (RC->Co[stage] != 0) || (RC->Ro[stage] != 0) )
       { Write_BS_Section(stage,RC,CF);}
       else
       {fprintf(CF,"\n*    Stage Number %d",stage+1);
        fprintf(CF,"\nR%d\t%d\t%d\t%8.3E",s+1,s+1,s+2,R);
        fprintf(CF,"\nR%d\t%d\t%d\t%8.3E",s+2,s+2,s+3,R);
        fprintf(CF,"\nC%d\t%d\t%d\t%8.3E",s+1,s+3,0,C);
        fprintf(CF,"\nC%d\t%d\t%d\t%8.3E",s+2,s+2,s+11,C);
        fprintf(CF,"\nRa%d\t%d\t%d\t%8.3E",t,s+4,0,Ra);
        fprintf(CF,"\nRb%d\t%d\t%d\t%8.3E",t,s+4,s+11,Rb);
        fprintf(CF,
            "\nX%d\t%d\t%d\t%d\tOPAMP",t,s+3,s+4,s+11);
       }
      }
   }
```

Listing 5.5 Write_LP_Section function.

5.8.4 Compiling and Running ANALOG

All of the files needed to successfully generate the ANALOG executable are contained in the \ANALOG directory on the software disk provided with this text. These include the ANALOG.C, GET_INFO.C, and UTILITY.C source files as well as the associated header (.H) files and ERRORNUM.H. Either a make file (.MAK) or a project file (.PRJ) will probably be required by the compiler. The make or project file should be set to include the three source modules listed above. No special requirements are placed on the compile and link process, so the *small* memory model should be sufficient for the compilation. A copy of the ANALOG executable is provided in the root directory of the software disk if the user does not want to recompile the files.

In order to see how ANALOG functions, we can use it to generate the component values and PSpice data file for the problem given in Example 5.5. We can copy ANALOG.EXE to the hard disk of our computer for faster operation and execute it from the DOS prompt by simply typing ANALOG. The first screen we see as we run ANALOG is a welcome screen, which will be removed as soon as we press a key. The next screen, as shown in Figure 5.15, solicits the name of the file containing the filter coefficients and then displays the filter parameters. We can assume that the filter specifications have already been entered into the FILTER program, and data files have been generated with a basename of EXAMPLE5. Notice that no extension should be entered, although drive and directory information can be provided if the data file is stored in another location.

```
******************** ANALOG - Implementation Program ********************
********************* Copyright (c) 1994  Les Thede *********************

 Enter the filename (without extension) of a data file generated by
 FILTER. A circuit analysis file with .CIR extension will be created.
 Please enter filename: EXAMPLE5

                Example 5-5 <> Inv Cheby Lowpass 5th
                Filter implementation is:    Analog
                Filter approximation is:     Inv. Chebyshev
                Filter selectivity is:       Lowpass
                Passband gain        (dB):   -1
                Stopband gain        (dB):   -60
                Passband frequency   (Hz):   100
                Stopband frequency   (Hz):   300
                Filter Length or Order:      5
```

Figure 5.15 ANALOG data file display screen.

The next screen, shown in Figure 5.16, asks for component values which will be common and parameters for the PSpice analysis frequency response. ANALOG then displays the component values calculated from the filter coefficients. The user is then provided an opportunity to implement another active filter.

```
******************** ANALOG - Implementation Program ********************
********************* Copyright (c) 1994  Les Thede *********************

 Enter common capacitor value (Farads): 1E-8
 Enter common R_A value (Ohms):         1E+4
 Enter analysis starting freq (Hz):     10
 Enter number of analysis decades:      4

       Example 5-5 <> Inv Cheby Lowpass 5th
       Stage    R Value       R_B Value          Special
       =====  ===========    ===========    =================
         0    1.1550e+005    0.0000e+000    1st order section
         1    5.0455e+004    2.9129e+004    Co = 2.6203e-008
         2    3.1183e+004    4.8634e+004    Co = 6.8600e-008
              Common Component Values
       =====================================
         C    1.0000e-008   R_A  1.0000e+004
         R_x  2.2943e+005   R_y  1.0456e+004
       Implement another active filter? (Y/N): N
```

Figure 5.16 ANALOG analysis specification screen.

ANALOG generates a circuit analysis data file which can be used for PSpice analysis as shown in Listing 5.6. Notice that the first stage is first-order with an R_B value of one ohm (most circuit analysis tools don't accept zero ohms), while the other two stages describe twin-tee notch filters. The original calculated values are shown in the listing and should be used in a test analysis of the circuit to determine if the specifications have been met. If the specifications are not met with these "ideal" values, either the op-amp model is not adequate or an error has been encountered in the design steps.

The next step in the testing of the active filter is to replace the ideal components with practical values. For this exercise we will use 1% resistor and 2% capacitor values. (The selected values are shown in italics and parentheses in the listing.) A worst case analysis can then be run on the circuit which involves adding tolerance information for the component values. (Check your documentation for the procedure necessary to run a worst case analysis using your circuit analysis package.) During the worst case analysis, sensitivities for each component are determined, and as the final step, each component is set to its worst case extreme. Figure 5.17 shows both the worst case and nominal response. If the two frequency responses are carefully compared, we would see that there is a +2.5 dB "bump" in the passband, and we lose over 2 dB in the stopband attenuation. These changes should represent the "worst" that can happen due to the unfortunate selection of the worst possible combination of components. A Monte Carlo analysis can also be run on the circuit to indicate the more typical extremes which might be encountered.

Depending on the nature of the application, we can live with the resulting variations, select even more precise (expensive) components, or redesign the filter to more stringent specifications than actually desired. This redesigned filter would then be able to vary to some degree while still satisfying the real specifications. However, this filter may also require a higher-order, which will add cost to the project.

```
Example 5-5 <> Inv Cheby Lowpass 5th
*      Specify input source:
Vs        11      0       AC      1.0     0.0
*      Stage Number 1
R11       11      12      1.155E+005              (1.15E+005)
C11       12      0       1.000E-008              (1.00E-008)
Rb1       13      21      1
X1        12      13      21      OPAMP
*      Stage Number 2
C21       21      22      1.000E-008              (1.00E-008)
```

```
C22     22    24    1.000E-008          (1.00E-008)
C23     23    0     2.000E-008          (2.00E-008)
R21     21    23    5.046E+004          (5.10E+004)
R22     23    24    5.046E+004          (5.10E+004)
R23     22    31    2.523E+004          (2.55E+004)
Ra2     25    0     1.000E+004          (1.00E+004)
Rb2     25    31    2.913E+004          (2.94E+004)
Co2     24    0     2.620E-008          (2.70E-008)
X2      24    25    31    OPAMP
*    Stage Number 3
C31     31    32    1.000E-008          (1.00E-008)
C32     32    34    1.000E-008          (1.00E-008)
C33     33    0     2.000E-008          (2.00E-008)
R31     31    33    3.118E+004          (3.09E+004)
R32     33    34    3.118E+004          (3.09E+004)
R33     32    41    1.559E+004          (1.54E+004)
Ra3     35    0     1.000E+004          (1.00E+004)
Rb3     35    41    4.863E+004          (4.87E+004)
Co3     34    0     6.860E-008          (6.80E-008)
X3      34    35    41    OPAMP
*    Voltage divider section
Rx      41    42    2.294E+005          (2.32E+004)
Ry      42    0     1.046E+004          (1.05E+004)
*    Sub-circuit model for op-amp
.SUBCKT       OPAMP       1     2     6
Rin     1     2     1E+008
E1      3     0     1     2     1E+003
Rx      3     4     1E+003
Cx      4     0     1E-009
E2      5     0     4     0     1E+003
Rout    5     6     1E000
.ENDS
*    Analysis modes
.AC     DEC   100   1.000E+001  1.000E+005
.PROBE
.END
```

Listing 5.6 Circuit analysis data file for Example 5.5.

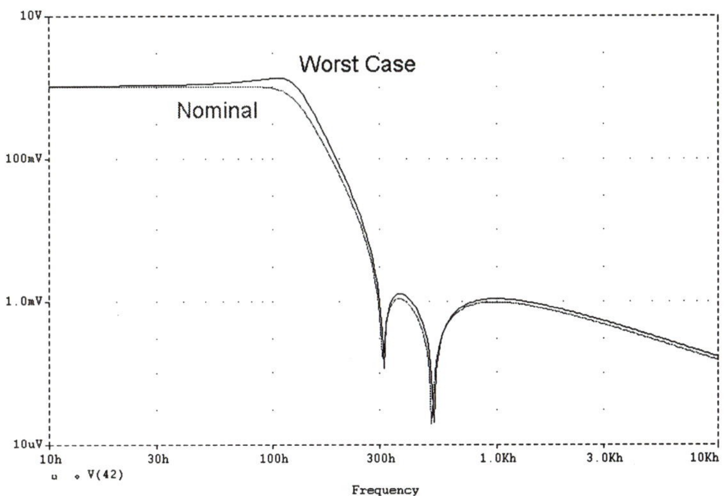

Figure 5.17 Frequency responses for Example 5.5.

5.9 CONCLUSION

We have reached the end of the first part of this text. We have been able to design a variety of analog filters, view their frequency responses, and implement them in an active filter form. Of course, we have left a good deal of material uncovered. There are other filter types which could have been discussed. There are other features which could have been included in the frequency response calculation and display. And there are other implementation techniques available for analog filters. These additions and embellishments can be added at a later time. We have tried to develop our C code in such a way that additional features can be added with a minimum of effort. But it's now time to move into the realm of digital filters. We will see that our work in the area of analog filters has not been wasted since one popular form of a digital filter uses much of what we have learned so far.

Chapter 6

Introduction to Discrete-Time Systems

Although most naturally occurring signals are of the analog variety (continuous-amplitude and continuous-time variation), we are finding that conversion of these signals to a digital form (discrete-time and discrete-amplitude variations) provides many advantages. For example, digital signals can be stored on computer floppy or hard disks. They can be compressed to save space, converted to other formats, or transmitted in combination with other signals. Digital forms of signals are truly becoming the standard in everyday use as compact disks (CDs) for audio and multimedia applications can attest. Therefore, the remainder of this text is devoted to the application of filtering techniques to digital signals.

The material in this chapter should provide a review of the basic principles of discrete-time systems which will be necessary to understand the material presented in the remainder of the text. In the first section of this chapter we will discuss the process of converting analog signals into a digital form. Next we will develop methods of dealing with discrete-time signals in both the time domain and frequency domain. We will also learn how to find the frequency response of a discrete-time system as well as how to play digitized waveforms on a computer equipped with a sound card.

6.1 Analog-to-Digital Conversion

As indicated in the introduction, most of the signals which we deal with every day are known as analog signals. This type of signal has a continuous variation in both time and amplitude as shown in Figure 6.1(a). In order to convert this analog signal to a digital signal with discrete-time and discrete-amplitude, as shown in Figure 6.1 (b), several steps must be performed.

First, the frequency spectrum of the analog signal must be strictly bandlimited. Second, the signal must be sampled at the proper sampling rate. And, third, the sampled value must be quantized to an acceptable level of accuracy. When it is time to convert the digital signal back to analog form, there are a number of methods which can be used, but one simple method requires only two steps. The first step is to output the digital value of the signal and hold it for the duration of the sample period. The second step is to pass that signal through a lowpass filter. In order to understand the reasons why these steps are necessary for analog and digital conversion, we will need to study the frequency spectrum of a sampled signal and the requirements placed on the sampling rate.

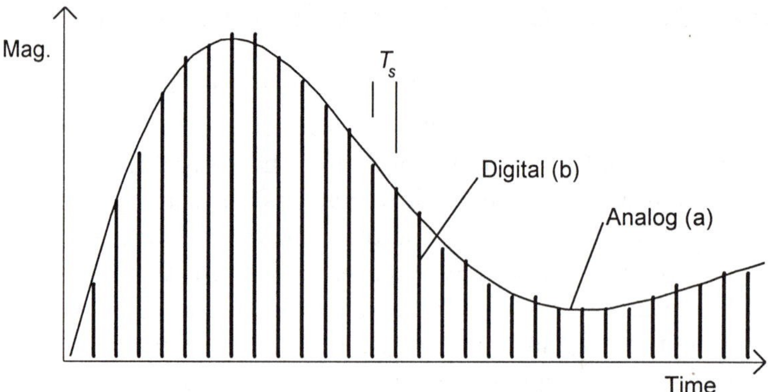

Figure 6.1 Comparison of (a) analog and (b) digital signals.

6.1.1 Frequency Spectrum and Sampling Rate

When an analog signal $x(t)$ is sampled as shown in Figure 6.1, the samples are usually taken at equal intervals of time. This sampling period T_S is the inverse of the sampling frequency f_S. The resulting digitized waveform $x_S(nT)$ can be specified with an argument indicating the sampling period T_S as shown in Equation 6.1.

$$x_s(nT) = x(t)\big|_{t=nT_s} \tag{6.1}$$

For example, if we had an analog signal $x(t)$ as specified in Equation 6.2, the sampled version of the signal $x_S(nT_S)$ as shown in Equation 6.3 would result.

$$x(t) = 50 \cdot e^{-100 \cdot t} \cdot \cos(200 \cdot t) \tag{6.2}$$

$$x(nT) = 50 \cdot e^{-100 \cdot nT_s} \cdot \cos(200 \cdot nT_s) \tag{6.3}$$

If we assume that the analog signal is sampled at a frequency of 1000 samples per second, then we can calculate the values of $x_S(nT_S)$ with T_S = 0.001 second. The values which result can be stored as a sequence of numbers as shown below. This is an important point: once an analog signal has been digitized, it is nothing more than a sequence of numbers which can be stored, manipulated, transmitted, or processed in any way we see fit.

$$x(nT_s) = \{50.0,\ 44.3,\ 37.7,\ 30.6,\ 23.4,\ 16.4,\ \ldots\},\ n = 0,\ 1,\ \mathrm{K} \tag{6.4}$$

Since the sampling period seldom changes, discrete-time equations usually drop the sampling period T_S from the expressions to produce an expression such as Equation 6.5. The sampling period will not be needed again until the digitized waveform is converted back to analog form.

$$x(n) = 25 \cdot e^{-10 \cdot n} \cdot \cos(100 \cdot n) \tag{6.5}$$

Although it doesn't appear that much has changed in the representation of the signal in the time domain, a great deal has changed in the frequency domain. The frequency spectrum of a signal is shown in Figure 6.2(a) before sampling. If this signal were sampled at a frequency of f_S, the spectrum of the sampled signal would be as shown in Figure 6.2(b). The original analog spectrum is replicated throughout the spectrum at intervals of f_S (although only one instance of that is shown). Because of this replication, it is feasible that there will be corruption of the frequency components of the original signal by components of the replicated signals. This corruption is referred to as *aliasing*, and the offending frequencies are *alias* frequencies. (Further details of aliasing and the upcoming Nyquist criteria can be found in most of the digital filter design references in Appendix A.)

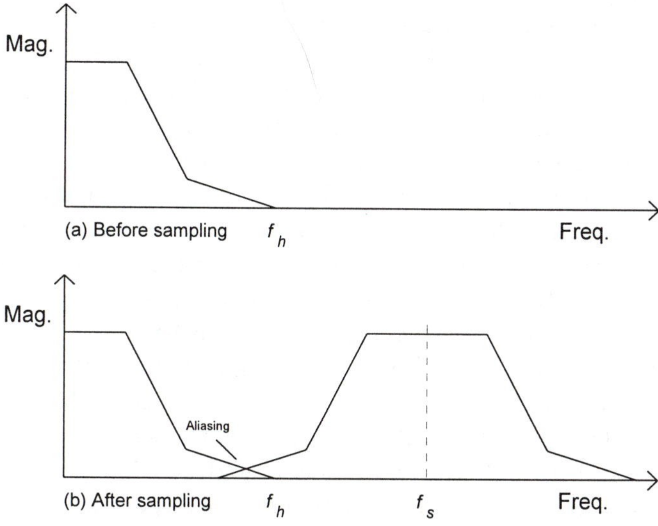

Figure 6.2 Spectrum of signal (a) before and (b) after sampling.

When digitizing a signal, it is very important to capture most, if not all, of the information present in the original analog signal without generating alias frequencies. For this reason, a good deal of study has gone into the conditions necessary to faithfully convert an analog signal into a digital form. We can see by closely observing Figure 6.2(b) that if we are to eliminate the effect of aliasing, the sampling frequency must be at least twice as high as the highest frequency in the original signal. This relationship, as shown in Equation 6.6, is known as the Nyquist criteria, and is a well-known requirement in sampling theory.

$$f_s > 2 \cdot f_h \tag{6.6}$$

In order to guarantee that this requirement is met at all times, it is normal procedure to bandlimit the input signal to one-half of the sampling frequency once the sampling frequency has been set. This is a prudent measure, since frequencies beyond those which are normally expected can occasionally occur in all systems. The bandlimiting process can be implemented by sending the analog signal through an analog lowpass filter prior to sampling. (What an excellent use of our analog filter theory!)

6.1.2 Quantization of Samples

Once the analog signal has been sampled, it is a discrete-time signal since values of the signal exist only at particular moments of time, but the amplitude of the signal is still continuous. Therefore, the next step in the analog-to-digital conversion (ADC) is to quantize the continuous-amplitude signal to one of many discrete values of amplitude. The number of possible values allowed for the amplitude is determined by the size of the variable chosen to store the values. For example, if a single byte of memory (8 bits) is chosen to store the information, then the amplitude can take on one of 2^8 or 256 different values. If the original signal had a range of amplitudes from +1 volt to −1 volt, then the difference between adjacent amplitudes would be approximately $7.8 \cdot 10^{-3}$ volts. On the other hand, if two bytes of memory (16 bits) were used to store each sample, there would be 2^{16} or 65,536 different values to represent the signal. With this many values, the ±1 volt signal would have adjacent amplitudes separated by only $3.05 \cdot 10^{-5}$ volts. Obviously, the larger the variable used to store the sampled data, the more closely we can approximate the analog signal with the digital representation.

However, the drawback of using larger and larger variables is twofold. First, the storage requirements to store the digitized waveform are proportional to the number of bits used to quantize the samples. For example, suppose we decide to sample a speech signal which contains signal frequencies from 300 – 3000 Hz at a frequency of 8000 Hz (which satisfies Nyquist's criteria). The waveform would need a file size of 480,000 bytes (480

Kbytes) to store one minute's worth of data using only 1 byte per sample. On the other hand, if we stored one minute of stereo music using 2 bytes of data for each channel (left and right), the file size would need to be over 10 million bytes (10 Mbytes). This large file size is necessary to accommodate a sampling rate of 44 kHz, which is the normal rate used for high-fidelity audio signal with frequencies up to 20 kHz.

The second drawback is the speed of conversion from analog-to-digital form. Although ADC chips are very fast these days, obtaining more accuracy requires more time for conversion. Eventually, a limit on accuracy will be reached because the conversion cannot be made in the allotted sample interval. This limitation is more common when processing video signals which have bandwidths in the millions of Hertz (MHz).

It is important to note at this point, that the sampling of an analog waveform does not normally produce any error. It is the quantization of the sample which produces the error in a digital system. If the samples could be stored in their original continuous-amplitude form, they could be used to regenerate the original signal with no error (assuming the Nyquist criteria is met). The maximum amount of error introduced into the system by this quantization is equal to one-half of the difference between amplitude levels. As we can see, selecting a method for the digitization of an analog signal is a compromise between conversion speed, waveform accuracy, and storage or transmission size.

6.1.3 A Complete Analog-to-Digital to Analog System

Figure 6.3 shows a block diagram of a complete system which first converts an analog signal to digital form for processing, transmission or storage. Then the conversion is undone by converting the digital signal back to analog form at another time and/or place using a digital-to-analog converter (DAC). As shown in Figure 6.2(b), the process of sampling a signal produces replicas of the original analog spectrum at intervals of the sampling frequency. In order to recover the original analog signal we simply have to eliminate the frequencies higher than $f_S/2$. This filtering is accomplished by a high-order lowpass filter.

A good example of such a complete procedure is the processing of audio signals for a music CD. The original sounds of the music are supplied by instruments or voices and are then recorded on tape in analog form. At some later time, these analog signals are digitized and encoded on the compact disk. We can then buy this CD and take it home and play it on our stereo system where the musical data is first converted from digital to analog form and then reproduced for us. In this example, the data is processed, stored and reproduced at a later time and different place.

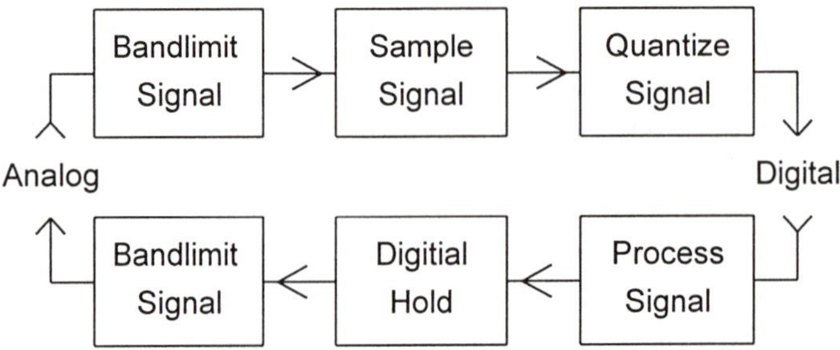

Figure 6.3 Complete analog-to-digital-to-analog system.

6.2 Linear Difference Equations and Convolution

In our study of discrete-time systems, we need to be able to describe them in a number of different ways. In this section we will learn how to describe a discrete-time system by using a difference equation and the system's impulse response. The impulse response of a system is simply its response to the discrete impulse function $\delta(n)$ as defined below.

$$\delta(n) = \begin{cases} 1, & \text{for } n = 0 \\ 0, & \text{for } n \neq 0 \end{cases} \tag{6.7}$$

In addition, we will be using the unit step function $u(n)$ in many of our expressions, so its definition is shown below as well.

$$u(n) = \begin{cases} 1, & \text{for } n \geq 0 \\ 0, & \text{for } n < 0 \end{cases} \tag{6.8}$$

6.2.1 Linear Difference Equations

One classic example of a discrete-time system which can be easily described by a difference equation is the bank account balance. Let $y(n)$ reflect the balance in a savings account where $x(n)$ dollars are deposited at the beginning of each month. We can assume that the account earns interest at a monthly percentage rate of I. If we let T_S, the sample period of the system, be one month, then the difference equation shown in Equation 6.9 reflects the balance in the account at the beginning of each month.

$$y(n) = y(n-1) + (I/100) \cdot y(n-1) + x(n) \tag{6.9}$$

If we assume that we deposit \$100 each month and the interest is 1% per month, we can replace $x(n)$ with $100 \cdot u(n)$ in Equation 6.9 to produce the following result.

$$y(n) = (1.01) \cdot y(n-1) + 100 \cdot u(n) \tag{6.10}$$

Table 6.1 shows the balance of the account for six months, assuming that we just opened the account and the balance was zero.

Table 6.1 Balance in savings account.

Month	Balance
0	\$100.00
1	\$201.00
2	\$303.01
3	\$406.04
4	\$510.10
5	\$615.20

Although manual techniques for the solution of this difference equation are acceptable for six months' time, other methods will need to be employed for longer periods of time. Let's see if we can determine a general formula for the balance of this account. Starting with the first month we find that

$$y(0) = 100 \tag{6.11}$$

$$y(1) = 1.01 \cdot y(0) + 100 = 100 \cdot (1 + 1.01) \tag{6.12}$$

$$y(2) = 1.01 \cdot y(1) + 100 = 100 \cdot (1 + 1.01 + 1.01^2) \tag{6.13}$$

or in general

$$y(n) = 100 \cdot \sum_{k=0}^{n} 1.01^k \tag{6.14}$$

Although this expression is compact, it still would require the calculation of the sum of $n + 1$ terms in order to determine the value. What we really need is a closed form solution. (Of course I wouldn't have mentioned such a thing if one didn't exist.) We can define what is referred to as a finite geometric sum as shown below.

$$\text{FGS} = \sum_{k=0}^{n} a^k \tag{6.15}$$

Then with some ingenious mathematics, we can define a difference which cancels most of the terms.

$$\text{FGS} - a \cdot \text{FGS} = \sum_{k=0}^{n} a^k - a \cdot \sum_{k=0}^{n} a^k = 1 - a^{n+1} \tag{6.16}$$

And, finally, we can determine the value of the FGS as shown below. (The value of FGS when $a = 1$ is determined directly from Equation 6.15.)

$$\text{FGS} = \begin{cases} \dfrac{1 - a^{n+1}}{1 - a}, & \text{for } a \neq 1 \\ n + 1, & \text{for } a = 1 \end{cases} \tag{6.17}$$

We can now write the equation for the balance of our savings account in a closed form as shown below. This expression does not need the calculation of $n + 1$ terms in order for us to evaluate it.

$$y(n) = 100 \cdot \frac{1 - 1.01^{n+1}}{1 - 1.01} \tag{6.18}$$

We can also define the value of the infinite geometric sum as shown below.

$$\text{IGS} = \sum_{k=0}^{\infty} a^k \tag{6.19}$$

By simply allowing n in Equation 6.17 to approach infinity, we can see that IGS can be specified as

$$\text{IGS} = \begin{cases} \dfrac{1}{1-a}, & \text{for } |a| < 1 \\ \text{undefined otherwise} \end{cases} \tag{6.20}$$

One way to completely define a discrete-time system is by using its difference equation. In general, the difference equation describing the output for a discrete-time system can be written as

$$y(n) = \sum_{k=0}^{M} a_k \cdot x(n-k) - \sum_{k=1}^{N} b_k \cdot y(n-k) \tag{6.21}$$

We would have complete knowledge of the system if we know the coefficients a_k and b_k. As indicated in Equation 6.21, the output $y(n)$ is a function of past and present values of the input $x(n)$ and past values of the output. A system which has its output described in terms of past values of the output is a *recursive* system. While a system which has its output described only by past and present values of the input is a *nonrecursive* system. We will see in the chapters to come, that these definitions effectively divide the types of digital filters to be designed into two groups as well.

Notice that we have defined the output of our system in terms of only past and present values of the input. Such a system is referred to as a *causal* system. Any real-time system of course must be causal since we can't determine the output of a system based on input or output values we haven't seen yet. However, systems which are not real-time can be *noncausal*. For example, any system which can draw its input from stored data can determine the output of the system at time n by using input values at $n + m$. These "future" values are known since they already have been stored.

6.2.2 Impulse Response and Convolution

Another way to completely describe a discrete system is to specify the impulse response $h(n)$ of the system. The impulse response of a system can be determined from the difference equation by substituting the impulse function $\delta(n)$ for the input $x(n)$ and determining the output $y(n)$.

Example 6.1 Determination of Impulse Response

Problem: Assume that a discrete-time system is described by the difference equation shown below. Determine the first five values of the impulse response, and then formulate an analytic expression for the impulse response.

$$y(n) = b_1 \cdot y(n-1) + x(n)$$

Solution: First, the equation is modified to reflect the fact that the output will be the impulse response $h(n)$ if the input is $\delta(n)$.

$$h(n) = b_1 \cdot h(n-1) + \delta(n)$$

Next, the set of the first five values of the impulse response are determined using the following table.

n	$\delta(n)$	$h(n-1)$	$h(n)$
0	1	0	1
1	0	1	b_1
2	0	b_1	b_1^2
3	0	b_1^2	b_1^3
4	0	b_1^3	b_1^4

From these results we can see that there is a general form to the impulse response as shown below.

$$h(n) = b_1^n \cdot u(n)$$

When the impulse response of a system is known, the complete characteristics of the system are known. The reaction of the system to any other input can then be determined by using convolution (similar to the operation in continuous systems). The output of the system can be defined as in Equation 6.22, where the * indicates convolution, not multiplication. In that expression, we see that the output of the system $y(n)$ is determined by computing a sum of products of the impulse response coefficients $h(n)$ and past values of the input $x(n-k)$. Convolution has the commutative property so the order of the functions in the convolution definition is not important. From a practical standpoint however, the simpler function is typically specified in the time-shifted format. This method is the predominant method used to implement an FIR filter which is discussed in detail in Section 9.3. In addition, a number of the texts listed in the digital filter design section of Appendix A cover discrete-time convolution.

$$y(n) = h(n) * x(n) = \sum_{k=-\infty}^{\infty} h(k) \cdot x(n-k) \tag{6.22}$$

Example 6.2 System Response by Convolution

Problem: Determine the output of the system described in Example 6.1 if the input is the signal shown below.

$$x(n) = a_1^n \cdot u(n)$$

Solution: Since we already have the impulse response of the system, we can use convolution to determine the output of the system. The equation for the output of the system is

$$y(n) = x(n) * h(n) = \sum_{k=-\infty}^{\infty} x(k) \cdot h(n-k)$$

or

$$y(n) = \sum_{k=-\infty}^{\infty} a_1^k \cdot u(k) \cdot b_1^{(n-k)} \cdot u(n-k) = b_1^n \cdot \sum_{k=0}^{n} (a_1/b_1)^k \cdot u(n)$$

Note that the limits on the convolution summation are adjusted by the functions $u(k)$ and $u(n-k)$. The step function $u(k)$ has zero value for negative values of k and therefore the lower limit of the summation is set to zero. The step function $u(n-k)$ will have zero value for all k values greater than n, and therefore the upper limit of the summation is set to n. We can use the form of a finite geometric sum to simplify the result into a closed form solution as shown below.

$$y(n) = \begin{cases} \dfrac{b_1^{n+1} - a_1^{n+1}}{b_1 - a_1}, & \text{for } b_1 \neq a_1 \\ (n+1) \cdot b_1^n, & \text{for } b_1 = a_1 \end{cases}$$

6.3 Discrete-Time Systems and z-Transforms

It is also important to be able to understand a discrete-time system's characteristics in the frequency domain as well as the time domain. For linear systems, the Laplace transform can be used to transform time domain characteristics to the frequency domain. For discrete-time systems, we will use the z-transform as defined in Equation 6.23.

$$z\{x(n)\} = X(z) = \sum_{n=-\infty}^{\infty} x(n) \cdot z^{-n} \tag{6.23}$$

The z-transform of a weighted impulse function, for example, results in a single term because the impulse function has only one nonzero value.

$$z\{A \cdot \delta(n)\} = A \sum_{n=-\infty}^{\infty} \delta(n) \cdot z^{-n} = A \cdot 1 \cdot z^0 = A \tag{6.24}$$

The z-transform of a weighted step function can also be determined using the definition. In this case, the result can be simplified by using the definition of the infinite geometric sum. Some of the more common z-transform pairs are shown in Table 6.2.

$$z\{A \cdot u(n)\} = A \sum_{n=-\infty}^{\infty} u(n) \cdot z^{-n} = A \cdot \sum_{n=0}^{\infty} (z^{-1})^n = \frac{A}{1 - z^{-1}} = \frac{A \cdot z}{z - 1} \tag{6.25}$$

The z-transform also has a set of useful properties as shown in Table 6.3. First and foremost is the property that the z-transform of the impulse response is the system's transfer function in the z-domain. We'll use that property in the next example. The second property in Table 6.3 shows that convolution in the time domain can be represented as simple multiplication

in the z-domain. The third property listed shows that time delay of k units of time can be represented by multiplication by z^{-k} in the z-domain. And finally, multiplication by n in the time domain can be represented by differentiation in the z-domain. For further discussion of these and other properties of the z-transform or for a more comprehensive list of z-transform pairs, please refer to one of the reference texts listed in Appendix A.

Table 6.2 Common z-transform pairs.

Time Domain Function	Frequency Domain Function
$A \cdot \delta(n)$	A
$A \cdot u(n)$	$\dfrac{A \cdot z}{(z-1)}$
$A \cdot n \cdot u(n)$	$\dfrac{A \cdot z}{(z-1)^2}$
$A \cdot a^n \cdot u(n)$	$\dfrac{A \cdot z}{(z-a)}$
$A \cdot \cos(\Omega \cdot n) \cdot u(n)$	$\dfrac{A \cdot z \cdot [z - \cos(\Omega)]}{z^2 - 2 \cdot \cos(\Omega) \cdot z + 1}$
$A \cdot \sin(\Omega \cdot n) \cdot u(n)$	$\dfrac{A \cdot z \cdot \sin(\Omega)}{z^2 - 2 \cdot \cos(\Omega) \cdot z + 1}$
$A \cdot a^n \cdot \cos(\Omega \cdot n + \phi) \cdot u(n)$	$\dfrac{A \cdot z \cdot [z \cdot \cos(\phi) - a \cdot \cos(\phi - \Omega)]}{z^2 - 2 \cdot a \cdot \cos(\Omega) \cdot z + a^2}$

Table 6.3 Common z-transform properties.

Time Domain Function	Frequency Domain Function
$h(n)$	$H(z)$
$\sum_{k=-\infty}^{\infty} x_1(k) \cdot x_2(n-k)$	$X_1(z) \cdot X_2(z)$
$x(n-k)$	$z^{-k} \cdot X(z)$
$n \cdot x(n)$	$-z \cdot \frac{dF(z)}{dz}$

Example 6.3 Determining the Transfer Function

Problem: Determine the transfer function of the system described in Example 6.1 which has an impulse response of

$$h(n) = b_1^n \cdot u(n)$$

Find the location of the poles and zeros of the transfer function as well as the difference equation of the system from the transfer function.

Solution: Using the fourth entry in the table of z-transform pairs, we can determine that the transfer function of the system described is

$$H(z) = \frac{z}{(z - b_1)} = \frac{1}{(1 - b_1 \cdot z^{-1})} = \frac{Y(z)}{X(z)}$$

We see that there is a single pole (denominator root) at $z = b_1$ and a single zero (numerator root) at $z = 0$.

By cross-multiplication, the following equation results.

$$Y(z) - b_1 \cdot z^{-1} \cdot Y(z) = X(z)$$

By taking the inverse z-transform of this equation and applying the time shift property of the z-transform, we can determine the same difference equation as in Example 6.1.

$$y(n) = b_1 \cdot y(n-1) + x(n)$$

Another way that a discrete-time system can be described is by drawing a system diagram to represent a general difference equation. Figure 6.4 shows the system diagram for Equation 6.21. In the figure, delays are represented by z^{-1} and multiplication by triangular symbols. Of course, if a nonrecursive filter was being represented, the lower half of the system diagram would be eliminated since the output of such a system does not depend on past values of the output.

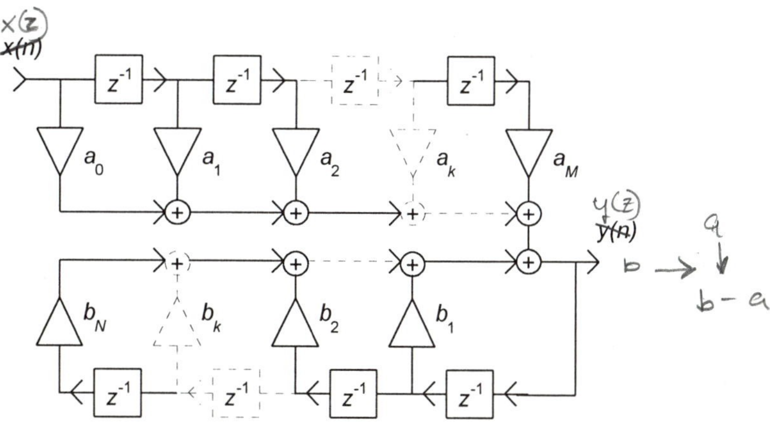

Figure 6.4 System diagram of general discrete-time system.

In this system diagram, we see that each time the signal moves through a delay element, the output is delayed by one sample interval. Although the system diagram is correct and will produce the correct output relationship, there is a more efficient description of the system. First, if we transform the general difference equation of Equation 6.21, we have

$$Y(z)\cdot\left(1+\sum_{k=1}^{N} b_k\cdot z^{-k}\right)=X(z)\cdot\sum_{k=0}^{M} a_k\cdot z^{-k} \tag{6.26}$$

From this description we can determine the transfer function $H(z)$ to be

$$H(z)=\frac{Y(z)}{X(z)}=\frac{\sum_{k=0}^{M} a_k\cdot z^{-k}}{1+\sum_{k=1}^{N} b_k\cdot z^{-k}} \tag{6.27}$$

We can represent Equation 6.27 in a slightly different way to allow for further development.

$$\frac{Y(z)}{W(z)}\cdot\frac{W(z)}{X(z)}=\frac{\sum_{k=0}^{M} a_k\cdot z^{-k}}{1}\cdot\frac{1}{1+\sum_{k=1}^{N} b_k\cdot z^{-k}} \tag{6.28}$$

This representation leads to two separate relationships as shown below.

$$Y(z)=W(z)\cdot\sum_{k=0}^{M} a_k\cdot z^{-k} \tag{6.29}$$

$$W(z)=X(z)+W(z)\cdot\sum_{k=1}^{N} b_k\cdot z^{-k} \tag{6.30}$$

These two equations in the z-domain can be written in their equivalent form in the time domain by recognizing that z^{-k} in the z-domain represents a delay of k sample periods in the time domain.

$$y(n)=\sum_{k=0}^{M} a_k\cdot w(n-k) \tag{6.31}$$

$$w(n)=x(n)+\sum_{k=1}^{M} b_k\cdot w(n-k) \tag{6.32}$$

The result of this derivation is that the general form of a discrete-time system diagram can be drawn with more efficient use of the delay units by defining w(n) in the diagram. Since these delay units must be implemented in hardware or software, the fewer used the better. Figure 6.5 shows the preferred method of drawing the system diagram with fewer delays.

6.4 Frequency Response of Discrete-Time Systems

The frequency response is one of the most important characteristics of a discrete-time system. Although it does not completely describe the system as the difference equation, impulse response or transfer function does (it does not convey the transient behavior of the system), the frequency response does provide important information about the steady-state behavior of the system. We can begin by considering an analog sinusoidal signal and the sampled signal using a sampling period of T_S.

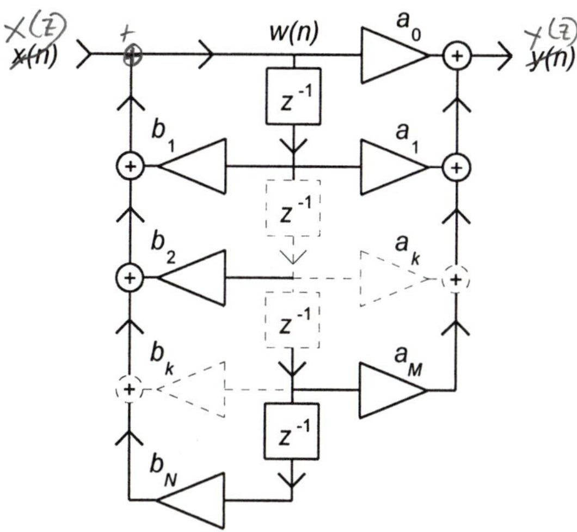

Figure 6.5 System diagram with fewer delay units.

$$x(t) = A \cdot \cos(\omega \cdot t) \tag{6.33}$$

$$x(nT) = A \cdot \cos(\omega \cdot nT_s) = A \cdot \cos(\Omega \cdot n) \tag{6.34}$$

where we have defined

$$\Omega = \omega \cdot T_s = 2\pi f \cdot T_s = 2\pi f / f_s \tag{6.35}$$

Equation 6.35 provides us with a method of comparing the analog frequency f to the digital frequency Ω. The range of analog frequencies acceptable for a discrete-time system does not extend from zero to infinity as would be the case in a normal analog system. We must remember that the upper limit of acceptable analog frequencies in discrete-time systems is governed by the Nyquist criteria, and therefore the range is defined as shown below. Note that a subscript of d is appended to the frequency variable to indicate we are talking about an equivalent analog frequency within a discrete-time system.

$$0 \le f_d < f_s / 2 \tag{6.36}$$

Combining this condition with Equation 6.35 results in the following range of digital frequencies

$$0 \le \Omega < \pi \tag{6.37}$$

Although a sine or cosine function is usually used to determine the frequency response of a hardware system, both of these functions can be described by complex exponentials as shown below.

$$\cos(\Omega \cdot n) = \frac{e^{j\Omega n} + e^{-j\Omega n}}{2} \tag{6.38}$$

$$\sin(\Omega \cdot n) = \frac{e^{j\Omega n} - e^{-j\Omega n}}{2j} \tag{6.39}$$

In fact, all periodic functions can be represented by these exponentials and even constants can be represented as exponentials with zero frequency. Therefore when determining the frequency response of a discrete system it is common to consider the driving function as the complex exponential as shown below.

$$x(n) = e^{j\Omega n} \tag{6.40}$$

The output of a discrete-time system can then be determined by convolving this input signal with the system's impulse response. The output is then

$$y(n) = \sum_{k=-\infty}^{\infty} h(k) \cdot e^{j\Omega(n-k)} = e^{j\Omega n} \sum_{k=-\infty}^{\infty} h(k) \cdot e^{-j\Omega k} \tag{6.41}$$

Or we can rewrite the result as

$$y(n) = x(n) \sum_{k=-\infty}^{\infty} h(k) \cdot e^{-j\Omega k} = x(n) \cdot H(e^{j\Omega}) \tag{6.42}$$

where $H(e^{j\Omega})$ is defined as the frequency response of the system.

If we compare the definition of the frequency response, as shown in Equation 6.43, and the definition of the transfer function for a system, as shown in Equation 6.44, we see a striking similarity.

$$H(e^{j\Omega}) = \sum_{k=-\infty}^{\infty} h(k) \cdot e^{-j\Omega k} \tag{6.43}$$

$$H(z) = \sum_{k=-\infty}^{\infty} h(k) \cdot z^{-k} \tag{6.44}$$

We can take advantage of this convenient similarity by defining the frequency response of a system in terms of the transfer function by simply allowing z to be replaced by $e^{j\Omega}$. That is,

$$H(e^{j\Omega}) = H(z)\big|_{z=e^{j\Omega}} \tag{6.45}$$

Example 6.4 Determining the Frequency Response

Problem: Determine the frequency response of the system described in Example 6.3 which has a transfer function of

$$H(z) = \frac{z}{(z - b_1)}$$

Assume that b_1 has a value of 0.8 for this system.

Solution: The frequency response can be found from the transfer function by simply making the substitution of $z = e^{j\Omega}$.

$$H(e^{j\Omega}) = H(z)\big|_{z=e^{j\Omega}} = \frac{e^{j\Omega}}{(e^{j\Omega} - 0.8)} = \frac{\cos(\Omega) + j \cdot \sin(\Omega)}{\cos(\Omega) - 0.8 + j \cdot \sin(\Omega)}$$

$$H(e^{j\Omega}) = \frac{1.0\angle\Omega}{\sqrt{(\cos(\Omega) - 0.8)^2 + (\sin(\Omega))^2}}$$

The values for the frequency response can then be calculated by allowing the frequency Ω to range from 0 to π as shown in the table below. Note that the complex exponential $e^{j\Omega}$ can be converted to a complex number in the rectangular form of $\cos(\Omega) + j\ \sin(\Omega)$ using Euler's relationship or converted to polar form as $1\angle\ \Omega$.

Frequency	Magnitude	Phase (deg)
0	5.00	0.00
$\pi/4$	1.42	–52.5
$\pi/2$	0.78	–38.7
$3\pi/4$	0.60	–19.9
π	0.56	0.00

Example 6.4 shows that the frequency response of a system is nothing more than a complex valued function of frequency. At any particular frequency, the complex value of the function can be determined and converted to polar form as shown below.

$$H(e^{j\Omega}) = M(\Omega)\angle\Phi(\Omega) \tag{6.46}$$

In this expression, $M(\Omega)$ represents the adjustment in magnitude that a signal will experience as it passes through the system, and $\Phi(\Omega)$ represents the adjustment in phase that the input signal will experience. For instance, if a digital signal

$$x(n) = 20\cos(\frac{\pi \cdot n}{2} + 30°) \tag{6.47}$$

was applied to the system of Example 6.4, the output signal could be determined by multiplying the magnitude of the input by the magnitude of the frequency response at $\Omega = \pi/2$, and by shifting the phase of the input by the phase of the frequency response at $\Omega = \pi/2$. The result could be written as

$$y(n) = 20 \cdot 0.78 \cdot \cos[(\pi \cdot n/2) + 30°–38.7°] \tag{6.48}$$

or

$$y(n) = 15.62 \cdot \cos[(\pi \cdot n/2) - 8.7°]. \tag{6.49}$$

Using a digital frequency of $\Omega = \pi/2$ may not be comfortable for analog filter designers who are used to working with much larger frequencies. Therefore it is important to note that this digital frequency does relate to some analog frequency through the formula expressed in Equation 6.35. That equation is restated here in terms of the equivalent analog frequency f_d. We see that the this frequency is a function of both the digital frequency and the sampling frequency for the system as discussed in Chapter 6.

$$f_d = \frac{\Omega}{2\pi} \cdot f_s \tag{6.50}$$

So if we are dealing with a system with a sampling frequency of 10,000 samples/second, it may be more natural to think of the input signal above as being an analog signal of 2500 Hz that has been converted to a digital frequency.

$$f_d = \frac{\pi / 2}{2\pi} \cdot 10{,}000 = 2500 \text{ Hz} \tag{6.51}$$

Working with discrete-time systems will require us to be able to use all of the material discussed thus far in this chapter. Some problems will be better expressed in the time domain, while others will be easier to work with in the frequency domain. Some system analysis or design will be easier using the system's impulse response, while others will require the use of the system's transfer function. As filter designers, we will definitely need to determine the frequency response of our system, and if we are to implement these filters, a system diagram will be handy. So before we leave this section, let's do an example to take in all the facets of discrete systems discussed so far.

Example 6.5 Complete Discrete-Time System Example

Problem: Consider the difference equation for a discrete-time system shown below. Determine the impulse response, transfer function, and frequency response of the system.

$$y(n) = 2.0 \cdot y(n-1) - 1.81 \cdot y(n-2) + 0.68 \cdot y(n-3) \\ + 1.0 \cdot x(n) + 3.0 \cdot x(n-1) + 3.0 \cdot x(n-2) + 1.0 \cdot x(n-3)$$

Also, find the location of the poles and zeros for the system as well as draw the system diagram. Finally, determine the output of the system if the input is the analog signal $x(t)$ shown below. Assume that the sampling frequency is 20,000 samples per second.

$$x(t) = 10 + 5\cos(2\pi \cdot 2000 \cdot t - 60°) + 20\sin(2\pi \cdot 8000 \cdot t + 30°)$$

Solution: We can begin the solution of the problem by transforming the difference equation in order to determine the transfer function for the system such that

$$Y(z) = 2.0 \cdot z^{-1} \cdot Y(z) - 1.81 \cdot z^{-2} \cdot Y(z) + 0.68 \cdot z^{-3} \cdot Y(z) + 1.0 \cdot X(z) \\ + 3.0 \cdot z^{-1} \cdot X(z) + 3.0 \cdot z^{-2} \cdot X(z) + 1.0 \cdot z^{-3} \cdot X(z)$$

or

$$\frac{Y(z)}{X(z)} = \frac{1.0 + 3.0 \cdot z^{-1} + 3.0 \cdot z^{-2} + 1.0 \cdot z^{-3}}{1.0 + 2.0 \cdot z^{-1} - 1.81 \cdot z^{-2} + 0.68 \cdot z^{-3}} = H(z)$$

After we convert $H(z)$ to positive powers of z, we factor the transfer function in order to determine the pole and zero locations. From the expression, we see that there are three zeros at $z = -1.0$ and three poles at $z = 0.8$ and $0.6 \pm j0.7$.

$$H(z) = \frac{(z + 1.0)^3}{(z - 0.8) \cdot (z^2 - 1.2 \cdot z + 0.85)}$$

The frequency response can now be easily determined from the transfer function as indicated below.

$$H(e^{j\Omega}) = \frac{(e^{j\Omega} + 1.0)^3}{(e^{j\Omega} - 0.8)\cdot(e^{j2\Omega} - 1.2\cdot e^{j\Omega} + 0.85)}$$

If we are to determine the response of our system to the sampled analog signal, we need to convert the critical analog frequencies to digital frequencies. The DC term is zero frequency in either case, while the analog frequency of 2000 Hz converts to a digital frequency of $\pi/5$, and the analog frequency of 8000 Hz converts to a digital frequency of $4\pi/5$. The magnitude and phase responses of $H(e^{j\Omega})$ are shown below.

Frequency	Magnitude	Phase (deg)
0.0	61.54	0
$\pi/5$	37.82	–88
$2\pi/5$	6.14	–249
$3\pi/5$	0.63	–261
$4\pi/5$	0.05	–266
π	0.00	–270

By applying the magnitude and phase adjustments indicated in the table for frequencies of 0, $\pi/5$, and $4\pi/5$, we can determine the output of the system as

$$y(t) = 615 + 189\cos(2\pi\cdot 2000\cdot t - 148^\circ) + 1.00\sin(2\pi\cdot 8000\cdot t - 236^\circ)$$

The impulse response of the system can be determined by finding the inverse z-transform of the transfer function. The most commonly used method for finding the inverse transform is by using partial fraction expansion of $H(z)/z$ and then matching the resulting terms to ones in the transform table.

$$\frac{H(z)}{z} = \frac{(z+1)^3}{z\cdot(z-0.8)\cdot(z^2 - 1.2\cdot z + 0.85)}$$

After using the standard methods for evaluation of coefficients, we find

$$\frac{H(z)}{z} = \frac{-1.4706}{z} + \frac{13.755}{(z-0.8)} - \frac{11.284 \cdot z - 7.5372}{(z^2 - 1.2 \cdot z + 0.85)}$$

We can then write $H(z)$ as the sum of three terms

$$H(z) = -1.4706 + \frac{13.755 \cdot z}{(z-0.8)} - \frac{11.284 \cdot z^2 - 7.5372 \cdot z}{(z^2 - 1.2 \cdot z + 0.85)}$$

The first two terms can be easily inverse transformed by matching terms in the z-transform table. However, the third term relates to a damped sinusoid and requires some manipulation before it can be inverse transformed.

$$\frac{11.284 \cdot z^2 - 7.5372 \cdot z}{(z^2 - 1.2 \cdot z + 0.85)} = \frac{A \cdot \cos(\phi) \cdot z^2 - A \cdot a \cdot \cos(\phi - \Omega) \cdot z}{z^2 - 2 \cdot a \cdot \cos(\Omega) \cdot z + a^2}$$

By comparing denominator terms, we find that $a = 0.92195$ and $\Omega = 0.86217$, while the numerator terms provide $A = 11.337$ and $\phi = 0.09677$. With these values, the transfer function can be inverse transformed to

$$h(n) = -1.4706 \cdot \delta(n) + 13.755 \cdot (0.8)^n \cdot u(n)$$
$$- 11.337 \cdot (0.92195)^n \cdot \cos(0.86217 \cdot n + 0.09677) \cdot u(n)$$

There was a good deal of algebra and trigonometry required to find the impulse response. If we need only a few of the first terms of the impulse response, or if we want to verify the correctness of our work there in a useful method which can be applied. If we return to the original $H(z)$ function and perform long division on the fraction, we will obtain a series of terms as shown.

$$H(z) = \frac{z^3 + 3.0 \cdot z^2 + 3.0 \cdot z + 1.0}{z^3 - 2.0 \cdot z^2 + 1.81 \cdot z - 0.68}$$

$$H(z) = 1.0 + 5.00 \cdot z^{-1} + 11.19 \cdot z^{-2} + 15.01 \cdot z^{-3} + \cdots$$

By inverse transforming this sequence, we have $h(n)$ represented as a series of delayed impulse functions. This will tell us explicitly what the value of the impulse response is for the first few values of the sequence. These values indeed check with those given by the general expression above for $h(n)$.

$$h(n) = \delta(n) + 5.00 \cdot \delta(n-1) + 11.19 \cdot \delta(n-2) + 15.01 \cdot \delta(n-3) + \cdots$$

6.5 Playing Digitized Waveforms on a Computer System

The addition of a variety of sound cards to computer systems is becoming an attractive option for many computer enthusiasts today. The sound cards add an extra dimension to many applications. Games are more fun to play with sound effects, word processing documents can have voice annotations, personal appointment calendars can provide voice reminders of upcoming meetings. These sound systems when combined with compact disk drives provide a total multimedia experience. Encyclopedias come to life with video and sound, wild animals can be admired and studied, and an explosion of other applications are available.

In order to get the full benefit of the work we will be doing in the remainder of this text, a computer sound card should be available. Certainly, the C code in this text for the design and implementation of analog and digital filters is the primary incentive for obtaining a copy of this text. But without a sound card to play the sound files which we will process in the next few chapters, an important experience will be missed. We will be using several sound files to illustrate the effects produced by the digital filters which we will be designing. These sound files have been included on the software disk with this text in the \SOUND directory.

There are a variety of sound file formats in use today. It seems that every manufacturer of sound equipment wants to "standardize" the industry. Two formats have been selected in which to provide sound files for this text in the hope that these formats will allow the widest application of the programs included with this text. The WAVE file format (.WAV) is used by Microsoft Corp. and a number of other companies to record and play sound files. The VOICE file format (.VOC) developed by Creative Labs, Inc., is also a very popular sound file format. In most cases, sound boards will either play these types of files or provide utilities to convert to and from these formats.

In addition to sound file formats, there are other variations which must be considered when playing sound files. First, the sampling frequency must be considered. The standard sampling frequency for studio-quality audio signals is 44,100 Hz. This frequency is high enough to allow audio frequencies in the 20,000 Hz range to be included in the signal information.

These frequencies represent the upper limit in the human hearing range. However, not all applications require this level of frequency response. Therefore sampling rates of 22,050 Hz and 11,025 Hz are also common. These lower frequencies provide attractive alternatives for signals without high frequency components or signals including only speech. The sound cards automatically include an antialiasing filter set to the correct frequency based on the sampling rate. Most sound cards also include the option of selecting the quantization to be used when the signal is sampled. Either 8-bit (1 byte) or 16-bit (2 bytes) resolution can be selected. Table 6.4 provides a look at the size of sound files as a function of sampling rate, number of channels and quantization method.

Table 6.4 Comparison of sound file size for 1 minute of recording.

Quantization	Channels	Sample Rate	File Size
8 bits	Mono (1)	11,025 Hz	0.662 MB
8 bits	Mono (1)	22,050 Hz	1.323 MB
8 bits	Mono (1)	44,100 Hz	2.646 MB
8 bits	Stereo (2)	11,025 Hz	1.313 MB
8 bits	Stereo (2)	22,050 Hz	2.646 MB
8 bits	Stereo (2)	44,100 Hz	5.292 MB
16 bits	Mono (1)	11,025 Hz	1.323 MB
16 bits	Mono (1)	22,050 Hz	2.646 MB
16 bits	Mono (1)	44,100 Hz	5.646 MB
16 bits	Stereo (2)	11,025 Hz	2.646 MB
16 bits	Stereo (2)	22,050 Hz	5.646 MB
16 bits	Stereo (2)	44,100 Hz	10.58 MB

Obviously, the size of audio files can grow very large! That is why it is important to select the sampling rate, number of channels and quantization method carefully to provide the level of accuracy appropriate to the project. There are two different files included with the software disk. Each sound file has been provided in both the .WAV and .VOC format. (Further details of these formats are provided in Section 9.4.1.) The first file, SPEECH, is a monaural file which uses a sampling rate of 11,025 samples per second with 8 bits per sample. The second selection, MUSIC, is also a monoaural signal recorded by sampling at 22,050 samples per second with 16 bits per sample. Since most sound cards will also record signals, other test signals can be captured and tested on the computer system as desired.

At this point, it is time to check out the sound card on the computer by playing the sound files mentioned above. (This may require finding and reading the documentation which came with the sound board!) It is recommended that the sound files be copied to the hard disk for faster access. If the sound board does not support playing the .WAV or .VOC files directly, check to see if there is a utility provided which will convert the files to a format supported by the sound board. Good luck!

6.6 CONCLUSION

We have reached the end of our review of discrete-time systems. In this chapter we found that the complete characteristics of a discrete-time system can be determined by knowing the system's difference equation, impulse response, transfer function or system diagram. We learned ways to determine any one of these descriptions from any of the others. In addition, we found out how to find the frequency response of a system by direct substitution into the system's transfer function. We will use the material presented in this chapter to design and implement digital filters in the remaining chapters. The remainder of this text will be presented in a manner emphasizing application rather than theory, but if the need arises, we can use the material in this chapter to better understand any problems which we might encounter.

Chapter 7

Infinite Impulse Response Digital Filter Design

There are a variety of methods which can be used to design digital filters as we will see in this chapter and the next. One commonly used method is to use the analog filter approximation functions which have already been developed and simply translate them in a way which will make them usable for discrete-time systems. This method, which will be studied in this chapter, makes use of the large backlog of filter design theory and tables of transfer functions which are readily available. Most of the filters designed using this method will be recursive in nature. That is, the output of the filter will depend on previous values of the output (as well as past and current values of the input). These types of filters can theoretically have impulse responses which continue forever and therefore are commonly referred to as infinite impulse response (IIR) filters.

Another method of designing discrete-time filters will be discussed in the next chapter. That method does not depend on analog filter theory, but rather uses the frequency response of the desired filter to directly determine the digital filter coefficients. The method generally yields nonrecursive filters which have outputs depending only on past and current values of the input. These types of filters generally have an impulse response containing only a finite number of values and thus are commonly called finite impulse response (FIR) filters. As we are about to see, both the IIR and FIR design methods will differ from the analog filter design techniques studied in the first part of the text. (A more complete comparison of IIR and FIR filters will be given in Section 9.1.)

In the first three sections of this chapter, we will investigate different methods of translating an analog filter's characteristics into those of a digital filter. As we will see, there is no perfect digital equivalent to an analog filter at all frequencies; however, we can develop filters which closely match the important filter characteristics. In the final sections of this chapter, we will develop and test the C code necessary to design IIR digital filters and to evaluate their frequency response characteristics.

7.1 Impulse Response Invariant Design

The impulse response invariant design method (or impulse invariant transformation) is based on creating a digital filter which has an impulse response which is a sampled version of the impulse response of the analog filter. We first start with an analog filter's transfer function $H(s)$, and by using the inverse Laplace transform, we determine the system's continuous impulse response $h(t)$. We next sample that response to determine the system's discrete-time impulse response $h(nT)$. We then take the z-transform of this sampled impulse response to find the discrete-time transfer function $H(z)$. As an illustration, consider the following example.

Example 7.1 Impulse Response Invariant Transformation

Problem: Assume that we wish to convert the following continuous-time transfer function to a discrete-time transfer function using the impulse invariant transformation method.

$$H(s) = \frac{12}{(s+2)(s+5)}$$

Solution: We first use basic partial fraction expansion techniques to write the transfer function in a form suitable for inverse transformation.

$$H(s) = \frac{4}{(s+2)} - \frac{4}{(s+5)}$$

Then recognizing the Laplace transform pair

$$L^{-1}\left\{ H(s) = \frac{A}{(s+a)} \right\} = A \cdot e^{-at} \cdot u(t) = h(t)$$

we can easily find the impulse response as

$$h(t) = (4 \cdot e^{-2t} - 4 \cdot e^{-5t}) \cdot u(t)$$

If we then sample this impulse response at intervals of T, we will have the discrete-time impulse response. Effectively, we simply replace every t with nT to denote the nth sample at intervals of T.

$$h(nT) = (4 \cdot e^{-2nT} - 4 \cdot e^{-5nT}) \cdot u(nT)$$

This expression can be rewritten in a form which more clearly indicates the exponential relationship of n.

$$h(nT) = [4 \cdot (e^{-2T})^n - 4 \cdot (e^{-5T})^n] \cdot u(nT)$$

Now we use the z-transform table as developed in Chapter 6 to find the transfer function in the z-domain.

$$H(z) = \frac{4}{1 - e^{-2T} z^{-1}} - \frac{4}{1 - e^{-5T} z^{-1}}$$

And, finally, we can combine the terms over a common denominator to produce the final result which can be simplified once a value of the sampling period T is chosen.

$$H(z) = \frac{4 \cdot (e^{-2T} - e^{-5T}) \cdot z^{-1}}{(1 - e^{-2T} z^{-1}) \cdot (1 - e^{-5T} z^{-1})}$$

Although Example 7.1 clearly indicates the steps required to translate an analog transfer function to a digital transfer function, we can skip some of the steps by recognizing the common relationship between $H(s)$ and $H(z)$. As can be verified in the example, for every term in the analog transfer function of the form

$$H(s) = \frac{1}{s + a} \tag{7.1}$$

there is a term created in the digital transfer function of the form

$$H(z) = \frac{1}{1 - e^{-aT} \cdot z^{-1}} \tag{7.2}$$

This matching technique can also be applied to quadratic terms which have complex roots, where each factor is simply treated individually. For example, if

$$H(s) = \frac{\beta}{(s + \alpha + j\beta)(s + \alpha - j\beta)} = \frac{j/2}{s + \alpha + j\beta} - \frac{j/2}{s + \alpha - j\beta} \tag{7.3}$$

the resulting discrete-time equivalent could then be written and simplified as follows.

$$H(z) = \frac{j/2}{1 - e^{-(\alpha + j\beta)T} \cdot z^{-1}} - \frac{j/2}{1 - e^{-(\alpha - j\beta)T} \cdot z^{-1}} \tag{7.4}$$

$$H(z) = \frac{(j/2) \cdot (e^{-(\alpha + j\beta)T} - e^{-(\alpha - j\beta)T})z^{-1}}{(1 - e^{-(\alpha + j\beta)T} \cdot z^{-1}) \cdot (1 - e^{-(\alpha - j\beta)T} \cdot z^{-1})} \tag{7.5}$$

$$H(z) = \frac{e^{-\alpha T} \cdot \sin(\beta T) \cdot z^{-1}}{1 - 2 \cdot e^{-\alpha T} \cdot \cos(\beta T) \cdot z^{-1} + e^{-2\alpha T} \cdot z^{-2}} \tag{7.6}$$

Example 7.2 Butterworth Impulse Invariant Filter Design

Problem: Determine the impulse invariant digital filter for a second-order Butterworth approximation function as shown below. Notice that $H(s)$ is normalized and therefore has a passband edge frequency of 1 rad/sec or ($1/2\pi$ Hz). Determine the differences which result from choosing sampling periods of $T = 1.0$ sec and $T = 0.1$ sec.

$$H(s) = \frac{1}{s^2 + 1.4142 \cdot s + 1}$$

Solution: We can first factor the analog transfer function and use partial fraction expansion to determine

$$H(s) = \frac{j0.7071}{s + 0.7071 + j0.7071} - \frac{j0.7071}{s + 0.7071 - j0.7071}$$

The digital transfer function can then be determined by using the results indicated in Equations 7.3 – 7.6.

$$H(z) = \frac{1.4142 \cdot e^{-0.7071 \cdot T} \cdot \sin(0.7071 \cdot T) \cdot z^{-1}}{1 - 2 \cdot e^{-0.7071 \cdot T} \cdot \cos(0.7071 \cdot T) \cdot z^{-1} + e^{-1.4142 \cdot T} \cdot z^{-2}}$$

We can now make the substitution of the different sampling periods in the general form to find the two distinct transfer functions.

$$H(z)\big|_{T=1.0} = \frac{0.45300 \cdot z^{-1}}{1 - 0.74971 \cdot z^{-1} + 0.24312 \cdot z^{-2}}$$

$$H(z)\big|_{T=0.1} = \frac{0.093096 \cdot z^{-1}}{1 - 1.85881 \cdot z^{-1} + 0.86812 \cdot z^{-2}}$$

It is interesting to compare the two transfer functions of Example 7.2. We notice first that the gains and pole positions are different solely from the selection of the sampling period (or frequency). We can also get a quick indication of the magnitudes of these transfer functions by determining the response at zero frequency. We can do that easily letting $z = e^{j0} = 1$.

$$H(e^{j0})\Big|_{T=1.0} = \frac{0.45300}{0.49341} = 0.91808$$

$$H(e^{j0})\Big|_{T=0.1} = \frac{9.3096 \cdot 10^{-2}}{9.3174 \cdot 10^{-3}} = 9.9917$$

With this quick check, we see that the response at zero frequency seems to be proportional to $1/T$. Although using only two values of sampling frequency hardly makes a case, it is true in general that the magnitude is proportional to $1/T$. For this reason, most impulse invariant designs scale the transfer function by an amount equal to the sampling period. As we can see, if that scaling were used in the previous example, the responses at zero frequency would be very close to unity.

The complete frequency responses for both transformations are shown in Figure 7.1 (with the T scaling factor applied). The responses are notably different as we would expect since the two transfer functions have different gains and pole locations. It is important to notice how this variation in transfer function form and frequency response is due solely to the value of sampling period (or sampling frequency) that has been selected.

In order to see why this selection produces the variations we must remind ourselves of the relationship between the digital and equivalent analog frequencies. As described in Section 6.4, the digital frequency Ω extends from 0 to π where π is analogous to the analog frequency of $f_s/2$ as dictated by the Nyquist criteria. Each point on the frequency axis can be referenced in terms of Ω, which extends from 0 to π, or in terms of f_d, which extends from 0 to $f_s/2$. This relationship can be written in either of the two forms shown in Equation 7.7 and 7.8.

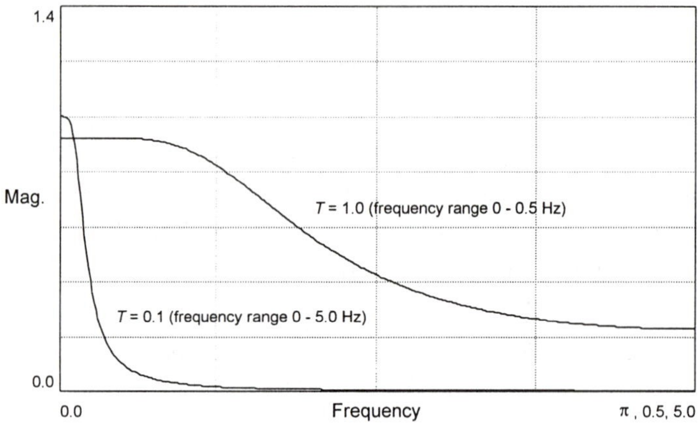

Figure 7.1 Frequency responses for Example 7.2.

$$\Omega = 2\pi f_d / f_s \tag{7.7}$$

$$f_d = f_s \frac{\Omega}{2\pi} \tag{7.8}$$

Now we are able to see more clearly why the responses are so different. Although the frequency axis extends from 0 to π, it represents different analog frequencies for the two responses. In the case of the $T = 1$ sec response (f_s = 1 Hz), the digital frequency range extends from 0 to 0.5 Hz, and the passband edge frequency of 0.159 Hz is clearly visible at a point approximately one-third of the way along the frequency axis. However, in the case of the $T = 0.1$ sec response (f_s = 10 Hz), the digital frequency range is actually from 0 to 5 Hz. Therefore, the break frequency of 0.159 Hz occurs at a point much closer to the zero frequency point.

Clearly then from this example it is important to pick the sampling frequency for an impulse invariant design carefully. The frequency range of the input signal must be considered as well as the desired overall response. In general, the impulse invariant design method is best for matching low frequency system responses.

7.2 Step Response Invariant Design

Another common method of converting an analog transfer function to the digital domain is to match the step response of both systems. The step response invariant design procedure is much the same as the impulse invariant design, except that we must determine the step response of the analog transfer function before it is sampled and z-transformed. We can determine the analog system's step response simply by multiplying $H(s)$ by the transform of the step input which is $1/s$. In Equation 7.9 we have defined the system's step response as $G(s)$.

$$G(s) = H(s) \cdot \frac{1}{s} \tag{7.9}$$

Once we have determined $G(s)$, we can find the time domain response to the step input $g(t)$ by using the inverse Laplace transform.

$$g(t) = L^{-1}\{G(s)\} \tag{7.10}$$

Then the discrete-time step response can be determined by sampling the continuous-time version.

$$g(nT) = g(t)\big|_{t=nT} \tag{7.11}$$

Next, the discrete-time system response to the step input can be determined by using the z-transform.

$$G(z) = Z\{g(nT)\} = H(z) \cdot \frac{1}{1 - z^{-1}} \tag{7.12}$$

As shown in Equation 7.12, $G(z)$ is the product of the discrete-time transfer function $H(z)$ and the z-transform of the step input. Therefore, in order to find $H(z)$, we simply multiply $G(z)$ by $(1 - z^{-1})$ as shown.

$$H(z) = G(z) \cdot (1 - z^{-1}) \tag{7.13}$$

As we can see as we review this procedure, the primary steps are the same as for the impulse invariant transformation, except that we have added one step at the beginning and one step at the end. The new initial step requires that we divide the analog transfer function by s, and the new final step requires that we multiply the digital transfer function by $(1 - z^{-1})$.

Example 7.3 Step Response Invariant Transformation

Problem: Assume that we wish to convert the continuous-time transfer function of Example 7.1 to a discrete-time transfer function, but this time we want to use the step response invariant transformation method.

$$H(s) = \frac{12}{(s+2)\cdot(s+5)}$$

Solution: We need to determine the system's response to a step input by multiplying $H(s)$ by the Laplace transform of the step input $1/s$.

$$G(s) = \frac{12}{s\cdot(s+2)\cdot(s+5)} = \frac{1.2}{s} - \frac{2.0}{s+2} + \frac{0.8}{s+5}$$

We can then easily find the time domain response to the step input by finding the inverse Laplace transform of $G(s)$ as

$$g(t) = (1.2 - 2.0\cdot e^{-2t} + 0.8\cdot e^{-5t})\cdot u(t)$$

If we sample this response at intervals of T, we will have the discrete-time response as shown below.

$$g(nT) = [1.2 - 2.0\cdot(e^{-2T})^n + 0.8\cdot(e^{-5T})^n]\cdot u(nT)$$

Then using the z-transform table in Chapter 6, we can find the z-transform of $g(nT)$ as

$$G(z) = \frac{1.2}{1 - z^{-1}} - \frac{2.0}{1 - e^{-2T}z^{-1}} + \frac{0.8}{1 - e^{-5T}z^{-1}}$$

And by combining the terms over a common denominator, we find that

$$G(z) = \frac{(1.2 - 2.0 \cdot e^{-2T} + 0.8 \cdot e^{-5T}) \cdot z^{-1} + (0.8 \cdot e^{-2T} - 2.0 \cdot e^{-5T} + 1.2 \cdot e^{-7T}) \cdot z^{-2}}{(1 - z^{-1}) \cdot (1 - e^{-2T} z^{-1}) \cdot (1 - e^{-5T} z^{-1})}$$

where $G(z)$ represents the output of the system to a step input. In order to determine the transfer function of the system, we must remove the effects of the step input $1 / (1 - z^{-1})$.

$$H(z) = \frac{(1.2 - 2.0 \cdot e^{-2T} + 0.8 \cdot e^{-5T}) \cdot z^{-1} + (0.8 \cdot e^{-2T} - 2.0 \cdot e^{-5T} + 1.2 \cdot e^{-7T}) \cdot z^{-2}}{(1 - e^{-2T} z^{-1}) \cdot (1 - e^{-5T} z^{-1})}$$

Although the result of the previous example looks quite involved, all of the exponential terms will become constants once the sampling period for the system is selected. We can also use the step invariant design method on the Butterworth filter of Example 7.2.

Example 7.4 Butterworth Step Invariant Filter Design

Problem: Determine the step invariant digital filter for a second-order normalized Butterworth approximation function as shown below. Determine the differences which result for sampling periods of $T = 1.0$ sec and $T = 0.1$ sec.

$$H(s) = \frac{1}{s^2 + 1.4142 \cdot s + 1}$$

Solution: Again, we first determine the output of the analog system to an input step function and then use partial fraction expansion to determine the individual terms.

$$G(s) = \frac{1}{s} - \frac{0.5 + j0.5}{s + 0.7071 + j0.7071} - \frac{0.5 - j0.5}{s + 0.7071 - j0.7071}$$

The step response can then be determined by finding the inverse Laplace transform of $G(s)$ as indicated below.

$$g(t) = 1.0 - (0.5 + j0.5) \cdot e^{-(0.7071 + j0.7071)t} - (0.5 - j0.5) \cdot e^{-(0.7071 - j0.7071)t}$$

Then after sampling at intervals of T, the discrete-time step response can be determined to be

$$g(nT) = 1.0 - (0.5 + j0.5) \cdot e^{-(0.7071 + j0.7071)nT} - (0.5 - j0.5) \cdot e^{-(0.7071 - j0.7071)nT}$$

The discrete-time response to the step input can then be determined by using the z-transform.

$$G(z) = \frac{1}{1 - z^{-1}} - \frac{0.5 + j0.5}{1 - e^{(0.7071 + j0.7071)T} z^{-1}} - \frac{0.5 - j0.5}{1 - e^{(0.7071 - j0.7071)T} z^{-1}}$$

Now by combining these terms over a common denominator (and performing a considerable amount of complex algebra), we have the system response to a step input.

$$G(z) = \frac{z^{-1} + e^{-0.707T}[(\sin(0.707T) - \cos(0.707T)]z^{-1}}{(1 - z^{-1}) \cdot [1 - 2e^{-0.707T}\cos(0.707T)z^{-1} + e^{-1.414T}z^{-2}]} + \frac{e^{-1.414T}z^{-2} - e^{-0.707T}[\sin(0.707T) + \cos(0.707T)]z^{-2}}{(1 - z^{-1}) \cdot [1 - 2e^{-0.707T}\cos(0.707T)z^{-1} + e^{-1.414T}z^{-2}]}$$

The discrete-time transfer function $H(z)$ can now be determined by removing the $(1 - z^{-1})$ factor relating to the step input, and we can make the substitution of the different sampling periods in the general form to find the two distinct transfer functions.

$$H(z) = \frac{z^{-1} + e^{-0.707T}[(\sin(0.707T) - \cos(0.707T)]z^{-1}}{[1 - 2e^{-0.707T}\cos(0.707T)z^{-1} + e^{-1.414T}z^{-2}]} + \frac{e^{-1.414T}z^{-2} - e^{-0.707T}[\sin(0.707T) + \cos(0.707T)]z^{-2}}{[1 - 2e^{-0.707T}\cos(0.707T)z^{-1} + e^{-1.414T}z^{-2}]}$$

$$H(z)\big|_{T=1.0} = \frac{0.94546 \cdot z^{-1} - 0.45205 \cdot z^{-2}}{1 - 0.74971 \cdot z^{-1} + 0.24312 \cdot z^{-2}}$$

$$H(z)\big|_{T=0.1} = \frac{0.13643 \cdot z^{-1} - 0.12711 \cdot z^{-2}}{1 - 1.8588 \cdot z^{-1} + 0.86812 \cdot z^{-2}}$$

Note that the pole locations are the same as for the impulse invariant design but that the zero locations have changed. We can compare the frequency responses of these two discrete-time filters as shown in Figure 7.2. There is significant difference between the two implementations, as we can see, but the differences are again a result of the frequency axis having two different scales. In the step invariant design method, there is no need to scale the magnitude as was the case for the impulse invariant design method. By comparing this frequency response to that of Figure 7.1, we see a significant difference. The reason for the difference is the different criteria placed on the design. In the previous section, the emphasis was placed on matching an impulselike input signal, while in this section the aim was to match a steplike input signal. As indicated in the figures, the different criteria produce filters with quite different frequency responses. As in the previous section, this method of IIR filter design is best suited to match low- frequency system responses.

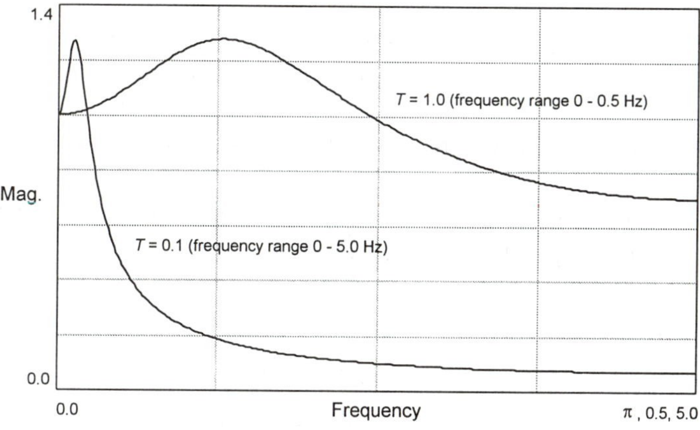

Figure 7.2 Frequency responses for Example 7.4.

7.3 Bilinear Transform Design

Both the impulse invariant and step invariant design methods provide good approximations for lowpass and some bandpass analog filter responses. However, they cannot provide good matching of high-frequency responses, which makes it impossible to use them for highpass or bandstop filter design. In fact, they do not provide the best methods for matching analog filter responses when a good match is required throughout a wide range of frequencies. In addition, without careful selection of the sampling frequency and strict bandlimiting of the input signal, distortion from aliasing can occur. Therefore, in this section we will discuss the bilinear transformation which endeavors to make a reasonable match over the entire filter frequency range. Of course, that provides a challenge since the analog frequency range extends from zero to infinity and the digital frequency range extends only from zero to π. However, a transformation from the analog s-domain to the digital z-domain has been developed (as described in more detail in the technical references provided at the end of the text). In this method, the relationship between the s and z complex variables can be described by the following equation where T is the sampling period.

$$s = \frac{2}{T}\frac{z-1}{z+1} \tag{7.14}$$

To better understand this relationship, we can represent the complex variable z in the exponential form $R \cdot e^{j\Omega}$

$$s = \frac{2}{T} \cdot \frac{R \cdot e^{j\Omega} - 1}{R \cdot e^{j\Omega} + 1} = \frac{2}{T} \cdot \frac{R \cdot \cos\Omega + j \cdot R \cdot \sin\Omega - 1}{R \cdot \cos\Omega + j \cdot R \cdot \sin\Omega + 1} \tag{7.15}$$

This representation can be written in rectangular form as

$$s = \frac{2}{T} \cdot \frac{[(R \cdot \cos\Omega - 1) + j \cdot R \cdot \sin\Omega] \cdot [(R \cdot \cos\Omega + 1) - j \cdot R \cdot \sin\Omega]}{[(R \cdot \cos\Omega + 1) + j \cdot R \cdot \sin\Omega] \cdot [(R \cdot \cos\Omega + 1) - j \cdot R \cdot \sin\Omega]} \tag{7.16}$$

and finally simplified to

$$s = \frac{2}{T} \cdot (\sigma + j\omega) = \frac{2}{T} \cdot \left[\frac{R^2 - 1}{R^2 + 2 \cdot R \cdot \cos\Omega + 1} + j \frac{2 \cdot R \cdot \sin\Omega}{R^2 + 2 \cdot R \cdot \cos\Omega + 1} \right] \tag{7.17}$$

By paying particular attention to Equation 7.17 and observing the s-plane and z-plane in Figure 7.3, we can see that there are three distinct regions in the s-domain which relate to three distinct regions in the z-domain. In the first case, any point in the z-domain which lies outside of the unit circle ($R > 1$) is associated with a point in the right-half plane (RHP) of the s-plane ($\sigma > 0$). In the second case, a point in the z-domain located inside the unit circle ($R < 1$) is associated with a point in the left-half plane (LHP) of the s-domain ($\sigma < 0$). And, finally, a point on the unit circle ($R = 1$) is associated with a point in the s-plane which lies on the $j\omega$ axis ($\sigma = 0$). In fact, in this last case, the positive $j\omega$ axis relates to the top half of the unit circle as the angle travels from 0 to π, while the negative $j\omega$ axis relates to the bottom half of the unit circle with angles from 0 to $-\pi$.

Although there does exist a one to one relationship between the positive $j\omega$ axis and the upper part of the unit circle, it is a nonlinear one. If we look more closely at the imaginary portion of Equation 7.17 when $R = 1$ (and therefore $\sigma = 0$), we see that

$$\omega = \frac{2}{T} \cdot \frac{\sin(\Omega)}{1 + \cos(\Omega)} = \frac{2}{T} \cdot \tan(\Omega / 2) \tag{7.18}$$

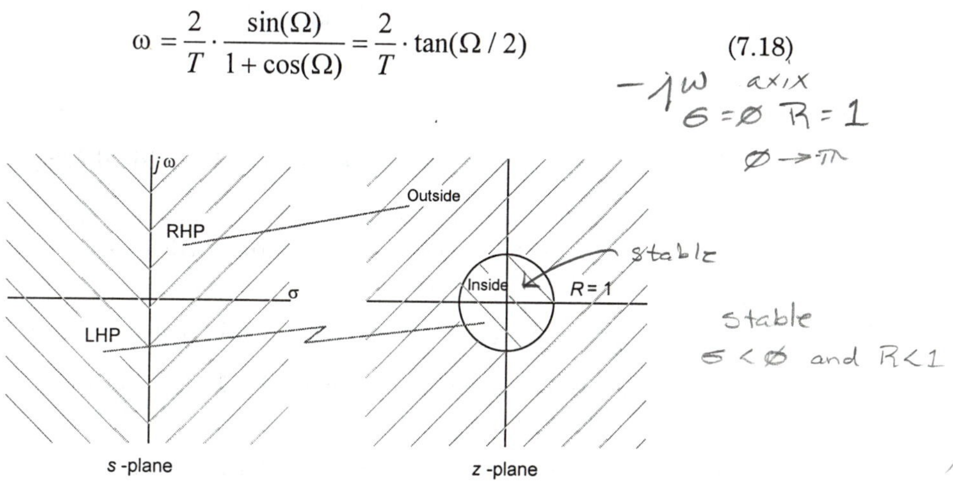

Figure 7.3 Comparison of s-plane and z-plane using bilinear transform.

or, in terms of the z-domain frequency variable,

$$\Omega = 2 \cdot \tan^{-1}\left(\frac{\omega T}{2}\right) \tag{7.19}$$

Equations 7.18 and 7.19 are important to the bilinear transformation process, because they are necessary to determine the proper mapping between the analog and digital domains. This mapping of analog frequencies to digital frequencies is fairly linear for low frequencies, but becomes very nonlinear as higher frequencies are mapped. This mapping of frequencies is often referred to as "warping" to describe how the higher frequencies are warped into their proper place on the unit circle. The reason that this warping is so important is that although we will specify the frequency characteristics of the digital filter by digital frequencies, the filter will be derived from an analog filter transfer function. Therefore it is necessary to properly determine the analog frequencies to use in the analog design by warping the specified digital frequencies as shown in Example 7.5.

After determining the frequencies necessary for the analog filter design, the filter can be designed using the process described earlier in this text. Once the analog transfer function has been determined, we can use the bilinear transform substitution given in Equation 7.14. Since we will be developing code to implement this transformation process, it is important to carefully describe this substitution. In the case of a first-order factor, the transformation process begins with Equation 7.20.

Example 7.5 Determining Analog and Digital Critical Frequencies

Problem: Assume we wish to design a lowpass digital filter (with sampling frequency of 20 kHz) based upon a Butterworth analog filter. The required characteristics of the *digital* filter are

$a_{pass} = -1$ dB, $a_{stop} = -20$ dB, $f_{pass} = 1$ kHz, and $f_{stop} = 5$ kHz

What are the parameters which we should use to design the *analog* filter upon which our digital filter will be based?

Solution: We first recognize that the digital frequency axis can be labeled in two different ways as discussed earlier. Then using Equation 7.7, we can determine

$$\Omega_p = \frac{2 \cdot \pi \cdot 1000}{20{,}000} = 0.1 \cdot \pi$$

$$\Omega_s = \frac{2 \cdot \pi \cdot 5000}{20{,}000} = 0.5 \cdot \pi$$

Once these frequencies have been determined they can be warped using Equation 7.18 to produce the equivalent analog frequencies necessary for the analog filter design.

$$\omega_p = \frac{2}{T} \cdot \tan\left(\frac{\Omega_p}{2}\right) = 2 \cdot \pi \cdot 1008.3 = 6335.4 \quad \text{rad/sec}$$

$$\omega_s = \frac{2}{T} \cdot \tan\left(\frac{\Omega_s}{2}\right) = 2 \cdot \pi \cdot 6366.2 = 40{,}000 \quad \text{rad/sec}$$

Note the slight warping of the lower passband edge frequency (from 1000 to 1008 Hz) and the more significant warping of the higher stopband edge frequency (from 5000 to 6366 Hz). The attenuations of the filter do not change; therefore, we have enough information to proceed with the design of the analog filter, which will be the subject of the next example.

$$H(z) = \left.\frac{A_1 \cdot s + A_2}{B_1 \cdot s + B_2}\right|_{s=\frac{2}{T}\cdot\frac{z-1}{z+1}} = \frac{A_1 \cdot \left(\frac{2}{T} \cdot \frac{z-1}{z+1}\right) + A_2}{B_1 \cdot \left(\frac{2}{T} \cdot \frac{z-1}{z+1}\right) + B_2} \tag{7.20}$$

In this equation, the uppercase *A*'s and *B*'s represent the coefficients of the analog filter function. After simplification, Equations 7.21 and 7.22 result with a new set of coefficients where 2 / *T* has been replaced with 2 * f_s. In those equations, the lowercase *a*'s and *b*'s represent the digital filter coefficients which will be used, and *G* represents the gain adjustment for this first-order term.

$$H(z) = \frac{N_0}{D_0} \cdot \frac{1 + \left(\frac{N_1}{N_0}\right) z^{-1}}{1 + \left(\frac{D_1}{D_0}\right) z^{-1}} = G \cdot \frac{a_0 + a_1 \cdot z^{-1}}{b_0 + b_1 \cdot z^{-1}} \tag{7.21}$$

where

$$\begin{aligned} N_0 &= A_2 + 2 f_s A_1 \\ N_1 &= A_2 - 2 f_s A_1 \\ D_0 &= B_2 + 2 f_s B_1 \\ D_1 &= B_2 - 2 f_s B_1 \end{aligned} \tag{7.22}$$

In the case of the quadratic terms that are used to describe our coefficients, the transformation process is shown in Equations 7.23 – 7.25. Again, *G* represents the gain adjustment necessary for each of the quadratic factors for the filter.

$$H(z) = \frac{A_0 \cdot \left(\frac{2}{T} \cdot \frac{z-1}{z+1}\right)^2 + A_1 \cdot \left(\frac{2}{T} \cdot \frac{z-1}{z+1}\right) + A_2}{B_0 \cdot \left(\frac{2}{T} \cdot \frac{z-1}{z+1}\right)^2 + B_1 \cdot \left(\frac{2}{T} \cdot \frac{z-1}{z+1}\right) + B_2} \tag{7.23}$$

$$H(z) = \frac{N_0}{D_0} \cdot \frac{1 + \left(\frac{N_1}{N_0}\right) z^{-1} + \left(\frac{N_2}{N_0}\right) z^{-2}}{1 + \left(\frac{D_1}{D_0}\right) z^{-1} + \left(\frac{D_2}{D_0}\right) z^{-2}} = G \cdot \frac{a_0 + a_1 \cdot z^{-1} + a_2 \cdot z^{-2}}{b_0 + b_1 \cdot z^{-1} + b_2 \cdot z^{-2}} \tag{7.24}$$

where

$$\begin{aligned}
N_0 &= A_2 + 2f_s A_1 + 4f_s^2 A_0 \\
N_1 &= 2 \cdot (A_2 - 4f_s^2 A_0) \\
N_2 &= A_2 - 2f_s A_1 + 4f_s^2 A_0 \\
D_0 &= B_2 + 2f_s B_1 + 4f_s^2 B_0 \\
D_1 &= 2 \cdot (B_2 - 4f_s^2 B_0) \\
D_2 &= B_2 - 2f_s B_1 + 4f_s^2 B_0
\end{aligned} \tag{7.25}$$

Example 7.6 Butterworth Bilinear Transform Filter Design

Problem: Determine the digital filter to meet the specifications given in Example 7.5 using the bilinear transformation.

Solution: By using the attenuations and the prewarped *analog* frequencies in the FILTER program designed earlier in the text, we determine the following analog transfer function.

$$H(s) = \frac{7.8877 \cdot 10^7}{s^2 + 1.2560 \cdot 10^4 s + 7.8877 \cdot 10^7}$$

Then by using the bilinear substitution for s, we can determine the transfer function in the digital domain.

$$H(z) = \frac{7.8877 \cdot 10^7}{\left(\frac{2}{T} \cdot \frac{z-1}{z+1}\right)^2 + 1.2560 \cdot 10^4 \cdot \left(\frac{2}{T} \cdot \frac{z-1}{z+1}\right) + 7.8877 \cdot 10^7}$$

The transfer function can be simplified by using Equations 7.23 – 7.25 to produce

$$H(z) = \frac{0.036161 \cdot (1 + 2 \cdot z^{-1} + z^{-2})}{(1 - 1.3947 \cdot z^{-1} + 0.53935 \cdot z^{-2})}$$

The frequency response of the digital filter designed in Example 7.6 can be determined in the manner discussed in the previous chapter. For a general quadratic factor of the form shown below,

$$H(z) = \frac{a_0 + a_1 \cdot z^{-1} + a_2 \cdot z^{-2}}{b_0 + b_1 \cdot z^{-1} + b_2 \cdot z^{-2}} \tag{7.26}$$

the frequency response can be determined by letting $z = e^{j\Omega}$ as shown.

$$H(e^{j\Omega}) = \frac{a_0 + a_1 \cdot e^{-j\Omega} + a_2 \cdot e^{-j2\Omega}}{b_0 + b_1 \cdot e^{-j\Omega} + b_2 \cdot e^{-j2\Omega}} \tag{7.27}$$

The numerator and denominator factors can then be converted. The frequency response of the filter designed in Example 7.6 is shown in Figure 7.4, where we can see that the specifications have been met (–1 dB gain = 0.89125 at 1000 Hz and –20 dB gain = 0.1 at 5000 Hz).

$$H(e^{j\Omega}) = \frac{[a_0 + a_1 \cos(\Omega) + a_2 \cos(2\Omega)] + j[a_1 \sin(\Omega) + a_2 \sin(2\Omega)]}{[b_0 + b_1 \cos(\Omega) + b_2 \cos(2\Omega)] + j[b_1 \sin(\Omega) + b_2 \sin(2\Omega)]} \tag{7.28}$$

7.4 C Code for Bilinear Transform IIR Filter Design

We have been developing our analog and digital filter design project since the beginning of this text by following an outline of the entire project. As indicated by the outline below, we are ready to implement item I.B.3

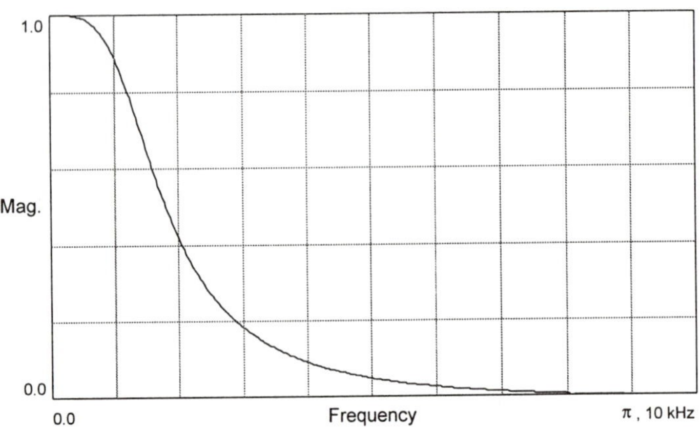

Figure 7.4 Frequency response for Example 7.6.

concerning the design of a digital IIR filter. We will be implementing this portion of our project design by using the bilinear transform approach to IIR filter design since it has wider application. However, there is no reason why the impulse or step invariance methods could not be implemented at a later time. All of the functions mentioned in this section can be found in the \FILTER\F_DESIGN.C module on the software disk provided with this text.

Analog and Digital Filter Design Project

- I. Design analog or digital filter.
  - A. Get filter specifications from user.
  - B. Design filter according to specifications.
    - 1. Design analog, if designated.
      - a. Calculate required order.
      - b. Calculate coefficients.
      - c. Unnormalize coefficients.
    - 2. Design digital FIR, if designated.
    - 3. Design digital IIR, if designated.
      - a. Prewarp critical frequencies.
      - b. Design analog filter.
      - c. Perform bilinear transformation.
      - d. Unwarp critical frequencies.
  - C. Display filter parameters on screen.
  - D. Save filter parameters to file.
- II. Determine frequency response of filter.
- III. Display magnitude and phase response on screen.

As shown in the project outline, our procedure will be the same as in the previous section: prewarp the critical digital frequencies to their analog counterparts, design a standard analog filter using the new specifications, perform the bilinear transform on the analog transfer function to produce the digital transfer function, and finally, set the frequencies back to their original values for future use. The design of the analog filter as indicated in items I.B.1.a–c has already been covered in earlier chapters, so we won't need to develop that code. (If we had been less careful in our project design, we might have had to rewrite some of the code in order to use it in the design of IIR filters.) Consequently, our work involves generating only a few new functions in order to add the IIR filter design capability to our project. So let's get started.

Our starting point in the IIR filter design process is the `Calc_Filter_Coefs` function as shown in Listing 7.1. This is where the process uses the implementation type to determine which coefficient calculation function to call. The next section will detail the `Calc_DigIIR_Coefs` function.

```
/*=======================================================
  Calc_Filter_Coefs() - determines implementation and
  calls appropriate calculation function
  Prototype:  int Calc_Filter_Coefs(Filt_Params *FP);
  Return:     error value
  Arguments:  FP - ptr to struct holding filter params
=======================================================*/
int Calc_Filter_Coefs(Filt_Params *FP)
{ int Error;                      /*  error value */

  /*  Call correct calc function for analog,
      digital FIR or IIR filters. */
  switch(FP->implem)
  { case 'A':
      Error = Calc_Analog_Coefs(FP);
      if(Error) { return 10*Error+1;}
      break;
    case 'F':
      Error = Calc_DigFIR_Coefs(FP);
      if(Error) { return 10*Error+2;}
      break;
    case 'I':
      Error = Calc_DigIIR_Coefs(FP);
      if(Error) { return 10*Error+3;}
```

```
        break;
      default:
        return ERR_FILTER;
    }
    return ERR_NONE;
}
```

Listing 7.1 Calc_FIlter_Coefs function.

7.4.1 Calc_DigIIR_Coefs Function

In the `Calc_DigIIR_Coefs` function shown in Listing 7.2, we see that the exact process described in our project outline is used. We first use `Warp_Freqs` to prewarp the digital frequencies; then, we sequentially call `Calc_Filter_Order`, `Calc_Normal_Coefs`, and `Unnormalize_Coefs` in order to design the analog filter. After the analog filter has been designed, the `Bilinear_Transform` function performs the conversion from analog coefficients to digital coefficients. And, finally, the `UnWarp_Freqs` function converts the critical frequencies back to their original values. Throughout this process, all critical parameters are transferred to and from the function in the `Filt_Params` structure using the pointer `FP`.

```
/*=======================================================
  Calc_DigIIR_Coefs() - calcs digital IIR coefs
  Prototype:  int Calc_DigIIR_Coefs(Filt_Params *FP);
  Return:     error value
  Arguments:  FP - ptr to struct holding filter params
=======================================================*/
int Calc_DigIIR_Coefs(Filt_Params *FP)
{ int Error;        /*  error value */

  /*  Pre-warp frequencies before making calcs */
  Error = Warp_Freqs(FP);
  if(Error) { return 10*Error+1;}
  /*  Calc order and coefs, then unnormalized coefs */
  Error = Calc_Filter_Order(FP);
  if(Error) { return 10*Error+2;}
  Error = Calc_Normal_Coefs(FP);
  if(Error) { return 10*Error+3;}
  Error = Unnormalize_Coefs(FP);
  if(Error) { return 10*Error+4;}
  /*  Transform from s-domain to z-domain */
  Error = Bilinear_Transform(FP);
```

```
    if(Error) { return 10*Error+5;}
    /*  Put the critical freqs back to orig value */
    Error = UnWarp_Freqs(FP);
    if(Error) { return 10*Error+6;}
    return ERR_NONE;
}
```

Listing 7.2 Calc_DigIIR_Coefs function.

7.4.2 Warp_Freqs and UnWarp_Freqs Functions

During the filter specification phase of our program, the user entered the desired frequency specifications which the digital filter must satisfy. In the `Warp_Freqs` function, these frequencies must be converted to analog frequencies using the techniques discussed earlier in this chapter. We can combine Equations 7.7 and 7.18 in order to determine the relationship which must exist between the analog and digital radian frequencies. The `UnWarp_Freqs` function simply resets the critical frequencies using the relationship of Equation 7.30.

$$\omega_a = \frac{2}{T} \cdot \tan\left(\frac{2 \cdot \pi \cdot f_d}{2 \cdot f_s}\right) = 2 \cdot f_s \cdot \tan\left(\frac{\omega_d}{2 \cdot f_s}\right) \tag{7.29}$$

$$\omega_d = 2 \cdot f_s \cdot \tan^{-1}\left(\frac{\omega_a}{2 \cdot f_s}\right) \tag{7.30}$$

7.4.3 Bilinear_Transform Function

```
/*=======================================================
  Bilinear_Transform() - use bilinear transform to
  convert transfer function from s-domain to z-domain
  Prototype:  int Bilinear_Transform(Filt_Params *FP,
                                     double fsamp);
  Return:     error value
  Arguments:  FP - ptr to struct holding filter params
              fsamp - sampling frequency (Hz)
=======================================================*/
```

```
int Bilinear_Transform(Filt_Params *FP)
{ int     i,j,start, /*  loop counters and index */
          num_quads; /*  number of quad factors */
  double  f2,f4,     /*  2 * fsamp, and 4 * fsamp^2 */
          N0,N1,N2,  /*  numerator temp variables */
          D0,D1,D2;  /*  denominator temp variables */
  if( (!FP->acoefs) || (!FP->bcoefs) )
  { return ERR_NULL;}
  /*  determine some constants  */
  f2 = 2 * FP->fsamp;
  f4 = f2 * f2;
  num_quads = (FP->order + 1)/2;
  /*  handle first-order factor if present  */
  /*  start indicates if one quad factor handled */
  start = 0;
  if(FP->order % 2)
  { N0 = FP->acoefs[2] + FP->acoefs[1] * f2;
    N1 = FP->acoefs[2] - FP->acoefs[1] * f2;
    D0 = FP->bcoefs[2] + FP->bcoefs[1] * f2;
    D1 = FP->bcoefs[2] - FP->bcoefs[1] * f2;
    FP->acoefs[0] = 1.0;
    FP->acoefs[1] = N1 / N0;
    FP->acoefs[2] = 0.0;
    FP->bcoefs[0] = 1.0;
    FP->bcoefs[1] = D1 / D0;
    FP->bcoefs[2] = 0.0;
    FP->gain *= (N0 / D0);
    start = 1;
  }
  /*  Handle quadratic factors. j indexes quad factors
      since each quad factor has three coefs  */
  for(i = start; i < num_quads ;i++)
  { j = 3 * i;
    N0 = FP->acoefs[j]*f4 + FP->acoefs[j+1]*f2 +
                                     FP->acoefs[j+2];
    N1 = 2 * (FP->acoefs[j+2] - FP->acoefs[j]*f4);
    N2 = FP->acoefs[j]*f4 - FP->acoefs[j+1]*f2 +
                                     FP->acoefs[j+2];
    D0 = FP->bcoefs[j]*f4 + FP->bcoefs[j+1]*f2 +
                                     FP->bcoefs[j+2];
    D1 = 2 * (FP->bcoefs[j+2] - FP->bcoefs[j]*f4);
    D2 = FP->bcoefs[j]*f4 - FP->bcoefs[j+1]*f2 +
                                     FP->bcoefs[j+2];
```

```
        FP->acoefs[j] = 1.0;
        FP->acoefs[j+1] = N1 / N0;
        FP->acoefs[j+2] = N2 / N0;
        FP->bcoefs[j] = 1.0;
        FP->bcoefs[j+1] = D1 / D0;
        FP->bcoefs[j+2] = D2 / D0;
        FP->gain *= (N0 / D0);
    }
    return ERR_NONE;
}
```

Listing 7.3 Bilinear_Transform function.

The objective of the `Bilinear_Transform` function shown in Listing 7.3 is to implement the transformation from an analog transfer function to a digital transfer function. A check that the `acoefs` and `bcoefs` arrays actually exist is made first in the function. Then constants which will be used often are calculated and stored in `f2` and `f4`. And, finally, the number of quadratics is calculated as well as a starting point for the quadratic loop. If the order of the filter is odd, then a first-order term is handled, and `start` is set to 1. Notice that `start` controls which quadratic factor will start the process. If a first-order factor (which is stored as a quadratic) has already been processed, the `for` loop will start with the second quadratic (if one is present). The actual transformation calculations are handled in exactly the same manner as derived in Equations 7.20 – 7.25. Also notice that the total gain of the filter is adjusted in the first-order case as well as in the quadratic loop. By the end of the function, all gain adjustments have been included and the analog coefficients have been replaced by digital IIR coefficients.

7.5 C CODE FOR IIR FREQUENCY RESPONSE CALCULATION

The next major activity of the program is to calculate the frequency response of the filter. The `Calc_Filter_Resp` function discussed in Chapter 4 will set up the necessary arrays and call the proper function based on the filter implementation. In this case, the `Calc_DigIIR_Resp` function shown in Listing 7.4 would be called to perform the frequency response calculations. Computations are made in the same manner as in `Calc_Analog_Resp` except the real and imaginary values are calculated using Equation 7.28.

```
/*=======================================================
  Calc_DigIIR_Resp() - calcs response for IIR filters
  Prototype:  int Calc_DigIIR_Resp(Filt_Params *FP,
                                   Resp_Params *RP);
  Return:     error value
  Arguments:  FP - ptr to struct holding filter params
              RP - ptr to struct holding respon params
=======================================================*/
int Calc_DigIIR_Resp(Filt_Params *FP,Resp_Params *RP)
{ int     c,f,q;          /*  loop counters */
  double  rad2deg,        /*  rad to deg conversion */
          omega,omega2,   /*  radian freq and square  */
          rea,img;        /*  real and imag part */

  rad2deg = 180.0 / PI; /*  set rad2deg */
  /*  Loop through each of the frequencies */
  for(f = 0 ;f < RP->tot_pts; f++)
  { /*  Initialize magna and angle */
    RP->magna[f] = FP->gain;
    RP->angle[f] = 0.0;
    /*  Pre calc omega and omega squared */
    omega = PI2 * RP->freq[f] / FP->fsamp;
    omega2 = 2 * omega;
    /*  Loop through coefs for each quadratic */
    for(q = 0 ;q < (FP->order+1)/2; q++)
    { /*  c is coef index = 3 * quad index */
      c = q * 3;
      /* Numerator values */
      rea = FP->acoefs[c] + FP->acoefs[c+1]*cos(omega)
                          + FP->acoefs[c+2]*cos(omega2);
      img = -FP->acoefs[c+1]*sin(omega)
                          - FP->acoefs[c+2]*sin(omega2);
      RP->magna[f] *= sqrt(rea*rea + img*img);
      RP->angle[f] += atan2(img,rea);
      /* Denominator values */
      rea = FP->bcoefs[c] + FP->bcoefs[c+1]*cos(omega)
                          + FP->bcoefs[c+2]*cos(omega2);
      img = -FP->bcoefs[c+1]*sin(omega)
                          - FP->bcoefs[c+2]*sin(omega2);
      RP->magna[f] /= sqrt(rea*rea + img*img);
      RP->angle[f] -= atan2(img,rea);
    }
    /* Convert to degrees */
```

```
      RP->angle[f] *= rad2deg;
    }
    /*  Convert magnitude response to dB if indicated */
    if(RP->mag_axis == LOG)
    { for(f = 0 ;f < RP->tot_pts; f++)
      { /* Handle very small numbers */
        if(RP->magna[f] < ZERO)
        { RP->magna[f] = ZERO;}
        RP->magna[f] = 20 * log10(RP->magna[f]);
      }
    }
    return ERR_NONE;
}
```

Listing 7.4 Calc_DigIIR_Resp function.

Example 7.7 Chebyshev Bilinear Transform Filter Design

Problem: Use FILTER to completely design a Chebyshev digital IIR filter and display the magnitude response. The specifications for this filter are

$a_{\text{pass}} = -1$ dB, $a_{\text{stop}} = -60$ dB, $f_{\text{pass}} = 1$ kHz, $f_{\text{stop}} = 5$ kHz, and

$f_{\text{samp}} = 50$ kHz

Solution: We can supply these values to FILTER as shown in Figure 7.5, and FILTER will calculate the digital IIR coefficients as shown in Figure 7.6. The magnitude response of the filter is shown in Figure 7.7.

In Figure 7.7 we can see the effect of sampling on the frequency response. In particular, we see the lowpass response replicated in mirror image form at the sampling frequency of 50 kHz. In addition, we see the lower half of the reflection at twice the sampling frequency of 100 kHz. If we had chosen a larger-frequency scale, we would see these replications reproduced at all multiples of the sampling frequency. Of course, in the typical discrete-time system, the antialiasing filter will be set to one-half of the sampling frequency (25 kHz in this case), and the user would not see the effects of the higher-frequency components.

```
********************** FILTER - The Design Program *********************
******************** Copyright (c) 1994  Les Thede ********************

Please select filter implementation:
       Analog (A)  Dig FIR (F)  Dig IIR (I)
Filter implementation: I

Please select filter selectivity:
       Lowpass (L)  Highpass (H)  Bandpass (P)  Bandstop (S)
Filter selectivity: L

Please select filter approximation:
       Butterworth (B)  Chebyshev (C)  Elliptic (E)  Inv-Cheby (I)
Filter Approximation: C

Please enter the sampling frequency (Hz): 50000
Please enter the passband gain (dB): -2
Please enter the stopband gain (dB): -60
Please enter the passband edge freq (Hz): 10000
Please enter the stopband edge freq (Hz): 20000

Please enter a title (40 char max):
Digital IIR Chebyshev Lowpass___________
```

Figure 7.5 Specification screen for FILTER.

```
                        Digital IIR Chebyshev Lowpass

                   Filter implementation is:     IIR (digital)
                   Filter approximation is:      Chebyshev
                   Filter selectivity is:        Lowpass
                   Passband gain        (dB):    -2.00
                   Stopband gain        (dB):    -60.00
                   Passband frequency   (Hz):    10000.00
                   Stopband frequency   (Hz):    20000.00
                   Sampling Frequency  (Hz):     50000.00

                              Order = 4
                       Overall Gain = 1.86714451e-002
       Numerator Coefficients               Denominator Coefficients
  [() +    z^-1       +    z^-2      ]  [() +    z^-1       +    z^-2       ]
  ==================================  ===================================
1 1.0  2.00000000e+00 1.00000000e+00  1.0 -6.20696688e-01  8.14430976e-01
2 1.0  2.00000000e+00 1.00000000e+00  1.0 -1.18935540e+00  5.04413209e-01
                         Press any key to continue
```

Figure 7.6 Coefficient values from FILTER.

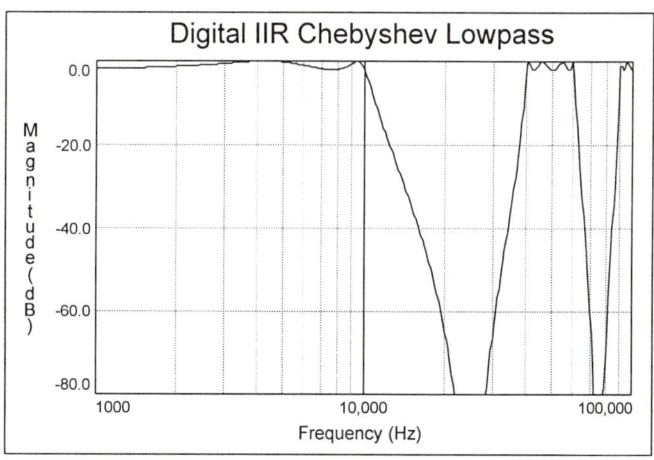

Figure 7.7 Magnitude response from FILTER.

7.6 CONCLUSION

In this chapter we have investigated three different methods of generating digital IIR filters from analog transfer functions. In each case we have found that there is no perfect match to the original analog function, primarily because there is a strict limit on the frequency range of a digital filter. However, the bilinear transform method does provide good overall response characteristics, and for that reason it was chosen to be implemented in C code. The frequency response characteristics of the IIR filter were also considered, and the calculation of the response was implemented in C code. And, finally, we used FILTER to design an IIR filter and display the complete magnitude response including the multiple replications of the original response. We are now ready to consider the design of FIR filters in the next chapter or the implementation of IIR filters in Chapter 9.

Chapter 8

Finite Impulse Response Digital Filter Design

In the last chapter, we considered the design of digital filters based on the approximation methods for analog filters. We investigated a number of ways that the transfer functions in the analog domain could be converted to transfer functions in the digital domain. In this chapter, we will develop methods which deal with the digital filter as a unique filter type, not based on analog filter approximation methods. The focus of this chapter will be on finite impulse response (FIR) filters which have only a finite number of terms in their impulse response. These filters have a number of advantages over the IIR filter types. An FIR filter is always stable, realizable, and provides a linear phase response under specific conditions. These characteristics make FIR filters attractive to many filter designers. However, the major disadvantage of FIR filters is that the number of coefficients needed to implement a specific filter is often much larger than for IIR designs. A more complete comparison of IIR and FIR filters is given in Section 9.1.

We will begin this chapter with a standard method of designing FIR digital filters using the Fourier series description of the desired frequency response. This method will then be modified and improved by using a windowing technique to improve the shape of the responses. In addition, the Parks-McClellan optimization technique will be discussed as a technique of reducing the length of the resultant FIR filters. And finally, the C code for designing and determining the frequency response of FIR filters will be developed.

8.1 Using Fourier Series in Filter Design

8.1.1 Frequency Response and Impulse Response Coefficients

In the process of filter design, the designer begins with the frequency response characteristics. The critical band edge frequencies and the gains within each band are determined to meet certain specifications. We have found that the frequency response for digital filters is actually periodic in the frequency domain with a period of the sampling frequency. For example, a typical lowpass filter specification is shown in Figure 8.1, which clearly indicates the periodic nature of the frequency response. Since this response is periodic, it can be described by a Fourier series of the form shown in

Equation 8.1. In this formulation, the complex frequency exponential is allowed to take on all possible frequency values.

$$H(e^{j\Omega}) = \sum_{k=-\infty}^{\infty} h(k) \cdot e^{-jk\Omega} \tag{8.1}$$

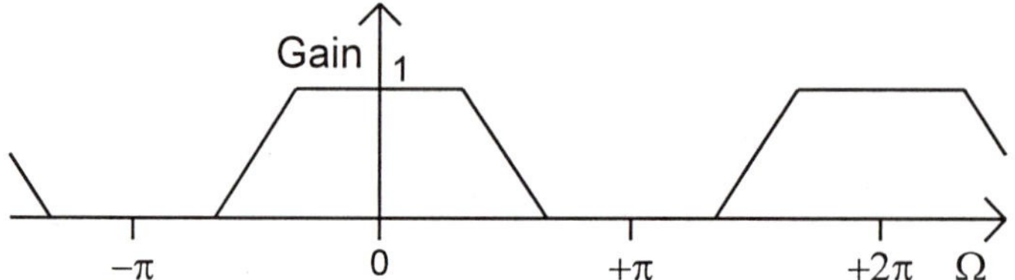

Figure 8.1 Periodic digital frequency response.

The coefficients within the summation are the impulse response coefficients which describe the digital FIR filter. The procedure for determining the impulse response coefficients from the frequency response is straightforward and provided in the digital filter design reference texts listed in Appendix A. The final result of the derivation is shown in Equation 8.2. As indicated by the limits of the integral, the integration must include only one full period of the frequency response.

$$h(n) = \frac{1}{2\pi} \int_{\Omega_o-\pi}^{\Omega_o+\pi} H(e^{j\Omega}) \cdot e^{jn\Omega} \cdot d\Omega, \quad n = 0, \ \pm 1, \ \pm 2, \ \text{K} \tag{8.2}$$

We will not be able to implement an infinite number of coefficients as Equation 8.2 indicates. The number of coefficients we retain is a compromise between how well we want our design to approximate the ideal, and how many coefficients can be retained because of time delay, implementation cost, or other constraints. We can assume that the indices are limited to the range $-M \le n \le +M$, which limits the number of coefficients retained to $N = 2M + 1$. By making this selection, we are in effect setting all other coefficients to zero. Figure 8.2 shows the effect of limiting the number of coefficients by graphing the frequency response using a finite number of coefficients. The frequency response can be determined by using a modified form of Equation 8.1 as shown in Equation 8.3.

$$H(e^{j\Omega}) = \sum_{n=-M}^{M} h(n) \cdot e^{-jn\Omega} \tag{8.3}$$

As we increase the number of coefficients in the FIR filter approximation, we can see that a ripple concentrates near the passband edge frequency. This ripple cannot be eliminated, even by increasing the number of impulse response coefficients; it simply concentrates at the transition. This effect is known as Gibbs's phenomenon and results whenever a discontinuity is modeled with a series. However, as we will see in the next section, there are methods we can use to reduce this effect.

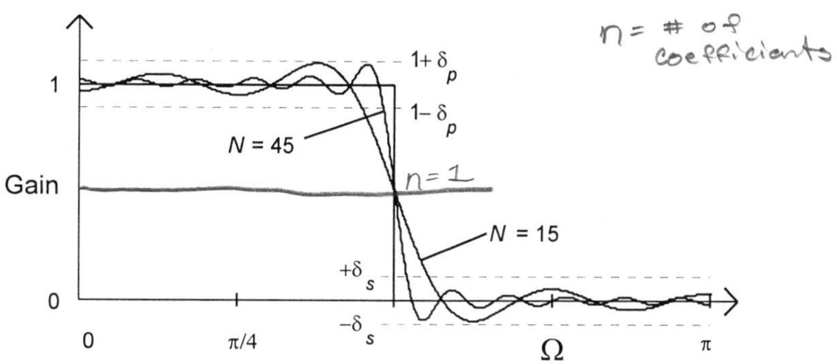

Figure 8.2 Approximated responses to lowpass filter.

Figure 8.2 also shows a more common method of specifying passband and stopband gain for FIR filters. The errors within the passband and stopband are specified as δ_p and δ_s, respectively. As we can see, the frequency response is allowed to fluctuate both positively and negatively within these error limits. We can translate these specifications into the decibel gain specifications with which we are familiar by using Equations 8.4 and 8.5. Alternatively, we can convert our decibel gains into these error values using Equations 8.6 and 8.7.

$$a_{\text{pass}} = 20\log(1-\delta_p) \tag{8.4}$$

$$a_{\text{stop}} = 20\log(\delta_s) \tag{8.5}$$

$$\delta_p = 1 - 10^{0.05 \cdot a_{\text{pass}}} \tag{8.6}$$

$$\delta_s = 10^{0.05 \cdot a_{\text{stop}}} \tag{8.7}$$

As indicated earlier, the phase response of an FIR filter can be a linear function of frequency under certain conditions. For example, if we assume that

$$H_{\text{passband}}(e^{j\Omega}) = 1 \cdot e^{-j\tau\Omega} = 1\angle - \tau\Omega \tag{8.8}$$

within the passband of an FIR filter, we are specifying that the gain must be unity, while the phase angle changes linearly with frequency. The necessary conditions which allow for this linear phase shift (or constant group delay) are that the impulse response coefficients be either symmetric or antisymmetric and that

$$\tau = \frac{N-1}{2} \tag{8.9}$$

where N is the number of coefficients or the length of the filter. The filter coefficients are symmetric if

$$h(n) = h(-n) \tag{8.10}$$

and they are antisymmetric if

$$h(n) = -h(-n) \tag{8.11}$$

Perhaps it is time to say a few words about the difference between filter length and filter order. Analog and digital IIR filters use the order of the filter as a measure of the filter's "size." The order refers to the highest-order term in the polynomial equation used to describe the filter. On the other hand, digital FIR filters typically use the number of impulse response coefficients required to describe it as its "size." This is probably because most FIR filters are implemented using convolution where the number of coefficients directly affects the length of the processing. Once the FIR coefficients are substituted into a difference equation (a little later in this section), we will see that the length of an FIR filter will simply be one larger than its order.

8.1.2 Characteristics of FIR Filters

When we consider symmetric and antisymmetric coefficients combined with even and odd filter lengths, four different types of FIR filters can be designed. Each of the four types has unique characteristics which can be described briefly as follows.

Type 1 FIR filters. The type 1 FIR filters, which have symmetric coefficients and odd length, also have a frequency response which has even symmetry about both $\Omega = 0$ and $\Omega = \pi$. This even symmetry allows the frequency response to take on any value at these two critical frequencies, and thus lowpass, highpass, bandpass, and bandstop filters can be implemented using this FIR type.

Type 2 FIR filters. The type 2 FIR filters, which have symmetric coefficients and even length, have a frequency response which is even about $\Omega = 0$ and odd about $\Omega = \pi$. This condition dictates that the response at $\Omega = \pi$ be zero and thus type 2 FIR filters are not recommended for highpass or bandstop filters.

Type 3 FIR filters. The type 3 FIR filters, which have antisymmetric coefficients and odd length, have a frequency response which has odd symmetry at both $\Omega = 0$ and $\Omega = \pi$. Because of the odd symmetry, the frequency response of this filter type must be zero at both of these two critical frequencies. Thus, this filter type is not recommended for lowpass, highpass or bandstop filters. However, this type of filter does provide a 90° phase shift of the output signal with respect to the input and therefore can be used to implement a differentiator or Hilbert transformer. This type of filter has other characteristics which make it the best choice for Hilbert transformation, while the differentiator is usually implemented using a type 4 filter.

Type 4 FIR filters. The type 4 FIR filters, which have antisymmetric coefficients and even length, have a frequency response which has odd symmetry about $\Omega = 0$ and even symmetry about $\Omega = \pi$. The odd symmetry condition make this type of filter a poor choice to implement either lowpass or bandstop filters. But, just as in the type 3 case, this filter provides a 90° phase shift which makes it able to implement differentiators and Hilbert transformers. This type of filter has better characteristics (in most cases) for implementing a differentiator than type 3, but the type 3 filter has some advantages over this filter type for implementing the Hilbert transform.

Since the type 1 FIR filter can be used to implement any of the filters we will need to design, we will be discussing only that type of filter from this point forward in this text. Further information concerning the other filter types can be found in a number of the digital filter design references listed in Appendix A.

The filter coefficients derived from Equation 8.2 will not produce a causal filter. This means that the system could not be implemented in real time. We can verify this if we consider the output of a discrete-time system produced by the convolution of the input signal with the impulse response coefficients as shown in Equation 8.12. Notice that the output $y(n)$ becomes only a function of the input $x(n)$ and does not include any past values of the output as in the IIR filter case.

$$y(n) = \sum_{k=-M}^{M} h(k) \cdot x(n-k) \tag{8.12}$$

As we see, using the impulse response coefficients directly will result in $y(n)$ being determined by future values of the input. For example, when $k = -M$, the summation includes a term $x(n + M)$ which refers to an input value M sampling periods ahead of $y(n)$'s reference time. The problem can be handled by simply shifting all coefficient values to the right on the time axis so that only positive values of n produce coefficients as shown in Figure 8.3. The only disadvantage of this action is to increase the time delay between system input and output by M sampling periods.

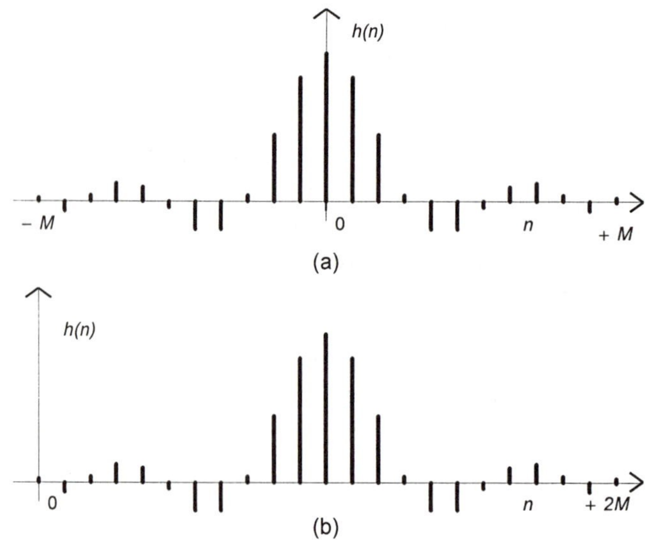

Figure 8.3 (a) Noncausal and (b) causal coefficients.

The causal coefficients can be determined from the noncausal coefficients by making the following index adjustments. As the noncausal coefficients indices take on values from $-M$ to $+M$, the causal coefficient indices will take on values from 0 to 2 M, as shown in Equation 8.13.

$$h_{\text{causal}}(n+M) = h_{\text{noncausal}}(n)\ ,\ n = 0,\ \pm 1, \mathrm{K}\ ,\ \pm M \tag{8.13}$$

8.1.3 Ideal FIR Impulse Response Coefficients

We can now determine the ideal coefficients for various filter types by using the integral formula of Equation 8.2. In each case, figures depicting ideal lowpass, highpass, bandpass, and bandstop filters along with equations for determining the coefficients based on the parameters of the particular filter are given. The frequency response in the passband of each filter is as defined in Equation 8.8 and allows us to determine causal coefficients directly. We will be assuming that the desired passband magnitude response is 1, while the stopband response is 0. We will be using δ_p and δ_s (or a_{pass} and a_{stop}) later to help determine the required length of the filter.

In the first case, Figure 8.4 illustrates the lowpass filter specification with the resulting derivation of the lowpass filter coefficients shown in Equations 8.14 and 8.15. The highpass, bandpass and bandstop filter cases are portrayed respectively in Figures 8.5 – 8.7 with the appropriate derivations in Equations 8.16 – 8.21. In each case, $\tau = M$ as determined from Equation 8.9.

Example 8.1 Determining Ideal Coefficients for an FIR Filter

Problem: Determine the ideal impulse response coefficients for a lowpass filter of length 21 to satisfy the following specifications:

$\omega_{\text{pass}} = 2\pi{\cdot}3000$ rad/sec, $\omega_{\text{stop}} = 2\pi{\cdot}4000$ rad/sec, and $f_s = 20$ kHz

Solution: We first need to determine Ω_c the cut-off frequency for the ideal filter. This frequency can be set in the middle of the transition band and converted to a digital frequency.

$$\Omega_c = (\omega_{\text{stop}} + \omega_{\text{pass}}) / (2 \cdot f_s) = 1.0996 \text{ rad/sec}$$

Using this value along with $\tau = 10$ in Equation 8.15, we can determine the following ideal causal coefficients.

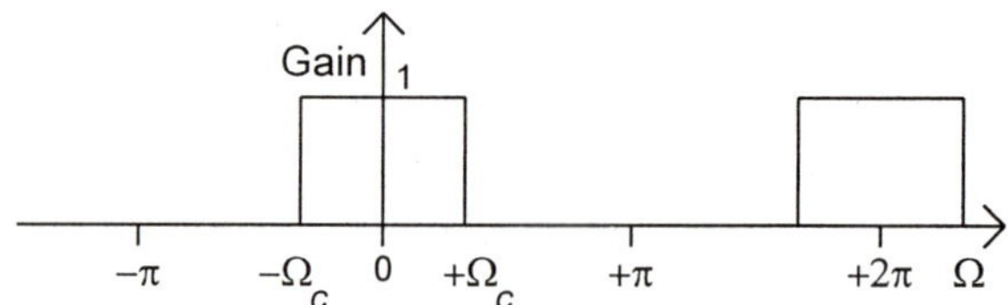

Figure 8.4 Lowpass filter specification.

$$h_{LP}(n) = \frac{1}{2\pi}\int_{-\Omega_c}^{+\Omega_c} e^{j(n-\tau)\Omega}\cdot d\Omega, \qquad (8.14)$$

$$n = 0,\ 1,\ \mathrm{K}\ ,\ 2M$$

$$h_{LP}(n) = \begin{cases} \dfrac{\sin[(n-\tau)\Omega_c]}{(n-\tau)\pi}, & \text{for } n \neq \tau \\ \Omega_c/\pi, & \text{for } n = \tau \end{cases} \qquad (8.15)$$

$$n = 0,\ 1,\ \mathrm{K}\ ,\ 2M$$

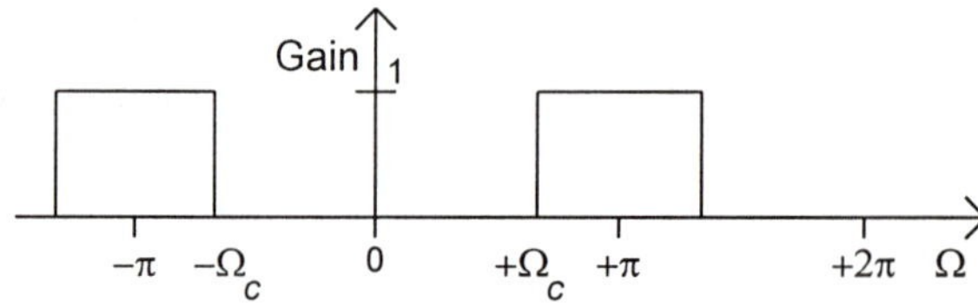

Figure 8.5 Highpass filter specification.

$$h_{HP}(n) = \frac{1}{2\pi}\left[\int_{-\pi}^{-\Omega_c} e^{j(n-\tau)\Omega}\cdot d\Omega + \int_{\Omega_c}^{+\pi} e^{j(n-\tau)\Omega}\cdot d\Omega\right] \qquad (8.16)$$

$$n = 0,\ 1,\ \mathrm{K}\ ,\ 2M$$

$$h_{HP}(n) = \begin{cases} \dfrac{\sin[(n-\tau)\pi] - \sin[(n-\tau)\Omega_c]}{(n-\tau)\pi}, & \text{for } n \neq \tau \\ (\pi - \Omega_c)/\pi, & \text{for } n = \tau \end{cases} \qquad (8.17)$$

$$n = 0,\ 1,\ \mathrm{K}\ ,\ 2M$$

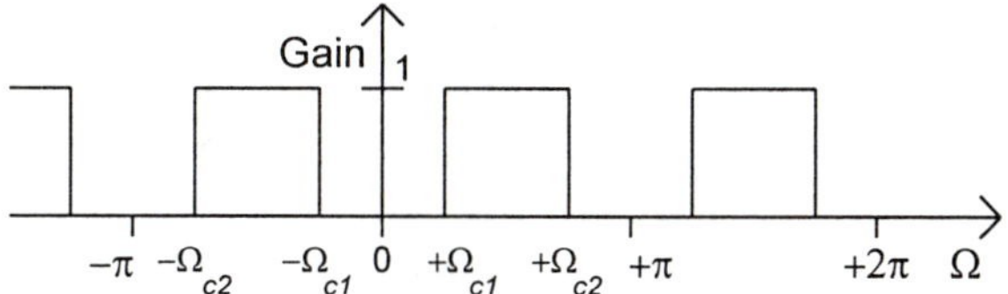

Figure 8.6 Bandpass filter specification.

$$h_{BP}(n) = \frac{1}{2\pi}\left[\int_{-\Omega_{c2}}^{-\Omega_{c1}} e^{j(n-\tau)\Omega} \cdot d\Omega + \int_{+\Omega_{c1}}^{+\Omega_{c2}} e^{j(n-\tau)\Omega} \cdot d\Omega\right] \quad (8.18)$$

$n = 0,\ 1,\ \mathrm{K}\ ,\ 2M$

$$h_{BP}(n) = \begin{cases} \dfrac{\sin[(n-\tau)\Omega_{c2}] - \sin[(n-\tau)\Omega_{c1}]}{(n-\tau)\pi}, & \text{for } n \neq \tau \\ (\Omega_{c2} - \Omega_{c1})/\pi, & \text{for } n = \tau \end{cases} \quad (8.19)$$

$n = 0,\ 1,\ \mathrm{K}\ ,\ 2M$

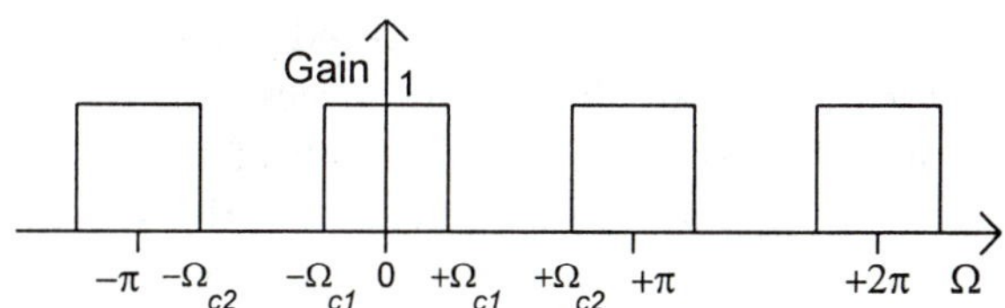

Figure 8.7 Bandstop filter specification.

$$h_{BS}(n) = \frac{1}{2\pi}\left[\int_{-\pi}^{-\Omega_{c2}} e^{j(n-\tau)\Omega} \cdot d\Omega + \int_{-\Omega_{c1}}^{+\Omega_{c1}} e^{j(n-\tau)\Omega} \cdot d\Omega + \int_{+\Omega_{c2}}^{+\pi} e^{j(n-\tau)\Omega} \cdot d\Omega\right] \quad (8.20)$$

$n = 0,\ 1,\ \mathrm{K}\ ,\ 2M$

$$h_{BS}(n) = \begin{cases} \dfrac{\sin[(n-\tau)\pi] - \sin[(n-\tau)\Omega_{c2}] + \sin[(n-\tau)\Omega_{c1}]}{(n-\tau)\pi}, & \text{for } n \neq \tau \\ (\pi - \Omega_{c2} + \Omega_{c1})/\pi, & \text{for } n = \tau \end{cases} \quad (8.21)$$

$n = 0,\ 1,\ \mathrm{K}\ ,\ 2M$

$$h(10) = 0.35000$$
$$h(9) = h(11) = 0.28362$$
$$h(8) = h(12) = 0.12876$$
$$h(7) = h(13) = -0.01660$$
$$h(6) = h(14) = -0.07568$$
$$h(5) = h(15) = -0.04502$$
$$h(4) = h(16) = 0.01639$$
$$h(3) = h(17) = 0.04491$$
$$h(2) = h(18) = 0.02339$$
$$h(1) = h(19) = -0.01606$$
$$h(0) = h(20) = -0.03183$$

8.2 Windowing Techniques to Improve Design

As indicated in the previous section, we are not able to include the infinite number of coefficients necessary to implement an ideal filter. We will have to reduce the number of coefficients used based on the constraints of our design. In the previous section, we simply truncated all noncausal coefficients beyond the indices $\pm M$ and kept the rest. (We will use the noncausal description of the filter coefficients for mathematical simplicity at this point. After the windowing process has been completed, we can shift the resulting coefficients to produce a causal filter.) This procedure can be compared to placing a window of width $N = 2M + 1$ over all of the ideal coefficients as shown in Figure 8.8. All of the coefficients within the window are retained and all coefficients outside of the window are discarded. In effect we have produced a rectangular "window" function in which all window coefficients with indices within the range of the window have a value of 1 and all other coefficients have a value of 0. The retained values of the filter coefficients would then be determined by performing a coefficient by coefficient multiplication of the ideal coefficients and the window coefficients as indicated in Equation 8.22.

$$h(n) = h_{\text{ideal}}(n) \cdot w(n),\ n = 0,\ \pm 1,\ \ldots,\ \pm M \tag{8.22}$$

The rectangular window coefficients can be formally defined in Equation 8.23. However, the abrupt truncation of the filter coefficients has an adverse effect on the resulting filter's frequency response. Therefore, a number of other window functions have been proposed which smoothly reduce the

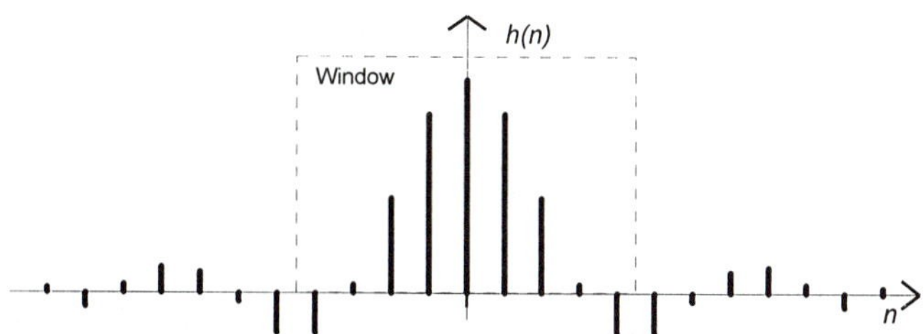

Figure 8.8 Window selection of coefficients

$$w_{\text{rect}}(n) = w_{\text{rect}}(-n) = 1, \; n = 0, \; 1, \; \text{K} \; , \; M \tag{8.23}$$

coefficients to zero. For example, a simple triangular window (also called the Bartlett window) as shown in Figure 8.9 would smooth the truncation process. An expression for these window coefficients is given in Equation 8.24 where $M = (N - 1) / 2$.

$$w_{\text{bart}}(n) = w_{\text{bart}}(-n) = \frac{(M-n)}{M}, \; n = 0, \; 1, \; \text{K} \; , \; M \tag{8.24}$$

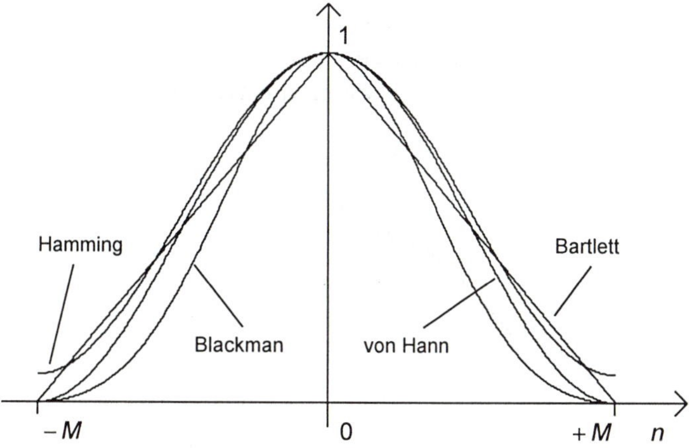

Figure 8.9 Bartlett, Blackman, Hamming, and von Hann windows.

Many window functions have been based on the raised cosine function including the von Hann, Hamming, and Blackman windows. Graphs of those windows are also shown in Figure 8.9 with the method of coefficient calculation for each window function provided in Equations 8.25 – 8.27.

$$w_{\text{hann}}(n) = w_{\text{hann}}(-n) = 0.5\left\{1 - \cos\left[\frac{\pi \cdot (M-n)}{M}\right]\right\}, \tag{8.25}$$

$n = 0,\ 1,\ K\ ,\ M$

$$w_{\text{hamm}}(n) = w_{\text{hamm}}(-n) = 0.54 - 0.46 \cdot \cos\left[\frac{\pi \cdot (M-n)}{M}\right], \tag{8.26}$$

$n = 0,\ 1,\ K\ ,\ M$

$$w_{\text{blck}}(n) = 0.42 - 0.5 \cdot \cos\left[\frac{\pi \cdot (M-n)}{M}\right] + 0.08 \cdot \cos\left[\frac{2\pi \cdot (M-n)}{M}\right] \tag{8.27}$$

and $w_{\text{blck}}(n) = w_{\text{blck}}(-n),\ n = 0,\ 1,\ K\ ,\ M$

As more time was spent trying to improve the window functions used in FIR filter design, it became apparent for a fixed length of filter that there was a trade-off between transition band roll-off and attenuation in the stopband. One of the window functions which developed because of this fact was the Kaiser window function, as shown in Figure 8.10. The expression for the window coefficients as given in Equation 8.28 is based on the modified Bessel function of the first kind I_0. The value β generally ranges from 3 to 9 and can be used to control the trade-off between the transition band and stopband characteristics.

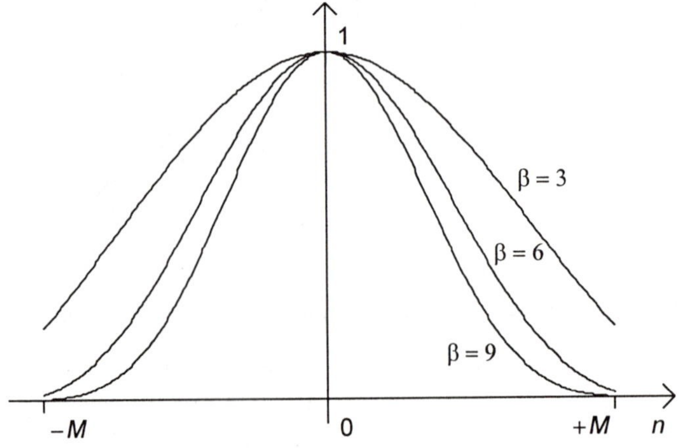

Figure 8.10 Kaiser windows with various β values.

$$w_{\text{kais}}(n) = w_{\text{kais}}(-n) = \frac{I_o\left[\beta \cdot \sqrt{1-\left(\frac{2 \cdot n}{M}\right)^2}\right]}{I_o(\beta)}, \tag{8.28}$$

$$n = 0,\ 1,\ \ldots,\ M$$

A reasonable estimate of β in the equation above has been determined empirically by Kaiser as shown in Equation 8.29.

$$\beta = \begin{cases} 0.1102 \cdot (A - 8.7), & \text{for } A > 50 \\ 0.5842 \cdot (A - 21)^{0.4} + 0.07886 \cdot (A - 21), & \text{for } 21 \le A \le 50 \\ 0.0, & \text{for } A < 21 \end{cases} \tag{8.29}$$

In Equation 8.29, the variable A represents the larger of the band errors (δ_p or δ_s) expressed as attenuation as shown in Equation 8.30.

$$A = -20 \cdot \log[\min(\delta_p, \delta_s)] \tag{8.30}$$

In addition, Kaiser developed empirical estimates of the filter length required to satisfy a given set of filter specifications as indicated in Equation 8.31.

$$N = \begin{cases} \dfrac{A - 7.95}{2.285 \cdot \Delta\Omega}, & \text{for } A > 21 \\ \dfrac{5.794}{\Delta\Omega}, & \text{for } A < 21 \end{cases} \tag{8.31}$$

In Equation 8.31, $\Delta\Omega$ represents the normalized radian transition band for lowpass and highpass filters and the smaller of the two normalized transition bands in the case of bandpass and bandstop filters.

$$\Delta\Omega = \left|\omega_{\text{stop}} - \omega_{\text{pass}}\right| / f_s \tag{8.32}$$

Once the desired window function has been selected, and the adjustments made to the ideal coefficients, the causal coefficients can be determined as indicated in the previous section.

Example 8.2 Determining Hamming Coefficients for an FIR Filter

Problem: Determine the coefficients for a lowpass filter using a Hamming window of length 21 to satisfy the specifications shown below:

$\omega_{pass} = 2\pi\cdot 3000$ rad/sec, $\omega_{stop} = 2\pi\cdot 4000$ rad/sec, and $f_S = 20$ kHz

Solution: The ideal coefficients have been determined in Example 8.1. We can use Equation 8.24 to determine the noncausal Hamming window coefficients as shown. After multiplication and shifting the coefficients by 10 sampling periods, the causal windowed coefficients result.

$w(0) = 1.0000$	$h(10) = 0.35000$
$w(1) = w(-1) = 0.97749$	$h(9) = h(11) = 0.27723$
$w(2) = w(-2) = 0.91215$	$h(8) = h(12) = 0.11745$
$w(3) = w(-3) = 0.81038$	$h(7) = h(13) = -0.01345$
$w(4) = w(-4) = 0.68215$	$h(6) = h(14) = -0.05163$
$w(5) = w(-5) = 0.54000$	$h(5) = h(15) = -0.02431$
$w(6) = w(-6) = 0.39785$	$h(4) = h(16) = 0.00652$
$w(7) = w(-7) = 0.26962$	$h(3) = h(17) = 0.01211$
$w(8) = w(-8) = 0.16785$	$h(2) = h(18) = 0.00393$
$w(9) = w(-9) = 0.10251$	$h(1) = h(19) = -0.00165$
$w(10) = w(-10) = 0.0800$	$h(0) = h(20) = -0.00255$

Figure 8.11 shows the frequency response for both Example 8.1 and 8.2. The rectangular window produces a filter which emphasizes transition band roll-off over ripple in the stopband. On the other hand, the filter produced by using Hamming coefficients has no noticeable ripple, but does not have a rapid roll-off in the transition band.

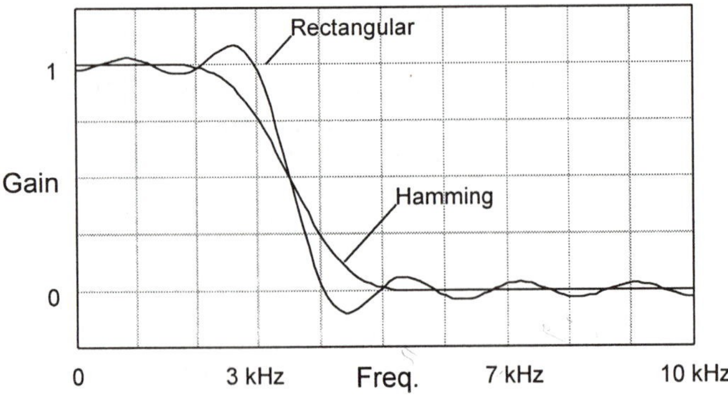

Figure 8.11 Magnitude responses for Examples 8.1 and 8.2.

Example 8.3 Determining Kaiser Coefficients for an FIR Filter

Problem: Determine the impulse response coefficients for a bandpass filter using a Kaiser window to satisfy the following specifications:

$f_{\text{pass1}} = 4$ kHz, $f_{\text{pass2}} = 5$ kHz, $f_{\text{stop1}} = 2$ kHz, $f_{\text{stop2}} = 8$ kHz, $a_{\text{pass1}} = -0.5$ dB, $a_{\text{stop1}} = a_{\text{stop2}} = -50$ dB, and $f_{\text{samp}} = 20$ kHz

Solution: The solution to this problem begins with the determination of the passband and stopband errors δ_p and δ_s by using Equations 8.6 and 8.7.

$$\delta_p = 1 - 10^{-0.025} = 0.055939$$

$$\delta_s = 10^{-2.5} = 0.0031623$$

We can then use Equations 8.29, 8.30, and 8.32 to find $A = 50$, $\beta = 4.53351$, and $\Delta\Omega$, where

$$\Delta\Omega_{\text{lower}} = \frac{2 \cdot \pi \cdot (4000 - 2000)}{20{,}000} = 0.6263185 \text{ rad/sec}$$

$$\Delta\Omega_{\text{upper}} = \frac{2 \cdot \pi \cdot (8000 - 5000)}{20{,}000} = 0.9424778 \text{ rad/sec}$$

Equation 8.31 can then be used to estimate of the smallest odd filter length as $N = 31$. Equation 8.17 can be used to determine the ideal filter coefficients after finding the values of $\tau = 15$ and calculating Ω_{c1} and Ω_{c2} as shown below.

$$\Omega_{c1} = \frac{2 \cdot \pi \cdot (2000 + 4000)}{2 \cdot 20{,}000} = 0.942478 \text{ rad/sec}$$

$$\Omega_{c2} = \frac{2 \cdot \pi \cdot (5000 + 8000)}{2 \cdot 20{,}000} = 2.042035 \text{ rad/sec}$$

The Kaiser window coefficients can be determined by using Equation 8.28. The final coefficients can be calculated by multiplying the ideal coefficients by the respective window coefficients and shifting all coefficient indices by a value of $M = 15$.

Rather than perform all of the numerical calculations by hand, we can use FILTER to finish the design of this filter. (We will discuss the C code for the FIR design procedure later in this chapter.) Figure 8.12 shows the FILTER input screen which indicates the filter parameters used to specify the filter. One addition to the screen for FIR filters is the indication of the filter length and an option to change the length if desired. Since the determination of filter length is not an exact calculation, this option allows the user to increase or decrease the length as necessary to satisfy the design.

```
********************** FILTER - The Design Program **********************
******************** Copyright (c) 1994  Les Thede *********************

Please select filter implementation:
       Analog (A)  Dig FIR (F)  Dig IIR (I)
Filter implementation: F

Please select filter selectivity:
       Lowpass (L)  Highpass (H)  Bandpass (P)  Bandstop (S)
Filter selectivity: P

Please select window or design method:
       Rectang (0)  Bartlett (1)  Blackman (2)   Hamming (3)
       Hanning (4)  Kaiser (5)    Parks-McCl (6):
Filter Window: 5

Please enter the sampling frequency (Hz): 20000
Please enter the passband gain (dB): -0.5
Please enter lower stopband gain (dB): -50
Please enter upper stopband gain (dB): -50
Please enter lower passband edge frq (Hz): 4000
Please enter upper passband edge frq (Hz): 5000
Please enter lower stopband edge frq (Hz): 2000
Please enter upper stopband edge frq (Hz): 8000

Please enter a title (40 char max):
Kaiser FIR Bandpass Filter

Filter length is estimated as 31.
Do you wish to change it? (Y/N) N
```

Figure 8.12 Specification screen for Example 8.3.

Figure 8.13 shows the FILTER coefficient screen with the final filter coefficients displayed. They are displayed in causal form, and are symmetric about the center value. The magnitude response of the filter is shown in Figure 8.14 and verifies that the specifications have been satisfied.

Kaiser FIR Bandpass Filter

```
Filter implementation is:          FIR (digital)
Filter approximation is:           Kaiser
Filter selectivity is:             Bandpass
Passband gain            (dB):     -0.50
Stopband gain - lower    (dB):     -50.00
Stopband gain - upper    (dB):     -50.00
Passband freq - lower    (Hz):     4000.00
Passband freq - upper    (Hz):     5000.00
Stopband freq - lower    (Hz):     2000.00
Stopband freq - upper    (Hz):     8000.00
Sampling Frequency (Hz):           20000.00

            Length = 31
     Overall Gain = 1.00000000e+000
```

Coefficients

N	[N + 0	N + 1	N + 2	N + 3]
0	-2.01200319e-003	-2.01615532e-003	4.85061935e-003	2.08877370e-003
4	2.97740954e-003	1.17871933e-002	-2.03738569e-002	-3.33459264e-002
8	1.95330713e-002	1.06179329e-002	1.48522841e-002	1.06004600e-001
12	-4.55615804e-002	-2.70313798e-001	2.58671685e-002	3.50000000e-001
16	2.58671685e-002	-2.70313798e-001	-4.55615804e-002	1.06004600e-001
20	1.48522841e-002	1.06179329e-002	1.95330713e-002	-3.33459264e-002
24	-2.03738569e-002	1.17871933e-002	2.97740954e-003	2.08877370e-003
28	-2.01200319e-003	4.85061935e-003	2.97740954e-003	

Figure 8.13 Coefficient values for Example 8.3.

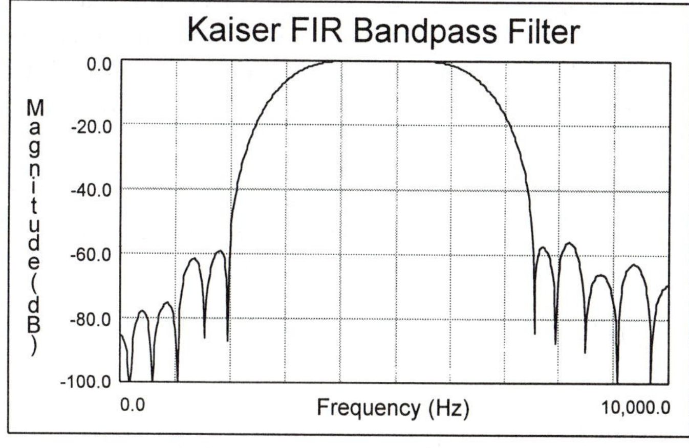

Figure 8.14 Magnitude response for Example 8.3.

8.3 Parks-McClellan Optimization Procedure

As we can see in Figure 8.14 for the filter using the Kaiser window, the stopband has ripple which generally decreases. In fact, if the passband characteristic were magnified, we would see ripple there as well. Both the passband and stopband ripple (error) tend to be larger near the transition bands and then taper off as the response moves away from the band edge. This type of response is not optimum. An optimum filter would have ripple in the passband and stopband which had a constant maximum magnitude. The error would still be present, but would be distributed equally throughout the bands. In this section, we will discuss a method for designing FIR filters with this characteristic. The Parks-McClellan algorithm, as it is generally known, was first presented over 20 years ago. Although the basic procedure has remained the same, a number of implementation techniques have changed. A detailed description of the Parks-McClellan (PM) algorithm is beyond the scope of this text, but a general overview of the procedure is appropriate. In addition, a review of the commented filter design code (as introduced in the next section) provides further implementation details. (Several of the texts listed in the digital filter design section of Appendix A provide more detailed descriptions of the PM algorithm, with the text by Antoniou providing a particularly detailed description.)

8.3.1 Description of the Problem

A primary component of the PM algorithm is a technique called the Remez exchange algorithm. But before we can we can use the power of the Remez algorithm to optimize our FIR filter coefficients, we must redefine our problem is such a way that the solution requires the minimization of an error function. To that end, we first define the frequency response of an odd-order FIR filter with symmetrical coefficients $h(n)$ as shown in Equation 8.33.

$$H(e^{j\omega}) = \sum_{n=0}^{N-1} h(n) \cdot e^{-j\omega n} = e^{-j\omega M} \sum_{m=0}^{M} c(m) \cdot \cos(m\omega) \tag{8.33}$$

where $M = (N-1)/2$ and

$$c(m) = \begin{cases} h(M), & \text{for } m = 0, \\ 2 \cdot h(M-m), & \text{for } m = 1, 2, \text{K} , M \end{cases} \tag{8.34}$$

We can then define the summation component of Equation 8.33 as

$$C(\omega) = \sum_{m=0}^{M} c(m) \cdot \cos(m\omega) \tag{8.35}$$

which can be used to formulate the error function which will be object of the minimization. As shown in Equation 8.36, the error function can be described in terms of the desired frequency response $e^{-j\omega M}\, D(\omega)$, the actual frequency response $e^{-j\omega M}\, C(\omega)$, and a weighting function $W(\omega)$ which can be used to adjust the amount of error in each filter band.

$$E(\omega) = W(\omega) \cdot [D(\omega) - C(\omega)] \tag{8.36}$$

The desired frequency response function $D(\omega)$ is usually defined as being 1 within the passband of the filter and 0 within the stopband, although other values can be assigned. The weighting function $W(\omega)$ can be defined equivalently throughout the filter band, or it can be assigned a value of 1 within the passband and 10 within the stopband if a smaller error value δ is desired in the stopband. This result occurs because the minimization algorithm will produce equal amounts of error throughout the defined frequency range, and since the stopband error has been artificially increased by 10, the actual error will be 10 times smaller.

The optimum error function will produce variations within the passband and stopband similar to those shown in Figure 8.2 (except that all ripple will be of the same magnitude). The actual error function will alternate between positive and negative δ values because of the summation of cosine functions. If we pick a set of frequencies ($x = M + 1$) at which the extremes of the error occur, Equation 8.36 can be written as

$$E(\omega_i) = W(\omega_i) \cdot [D(\omega_i) - C(\omega_i)] = (-1)^i \delta, \quad \text{for } i = 0, 1, \ldots, x \tag{8.37}$$

Equation 8.37 can be expanded into a matrix equation by considering these $x + 1$ frequencies which are typically called extremals and play a crucial role in the optimization process.

$$\begin{bmatrix} 1 & \cos\omega_0 & \cdots & \cos M\omega_0 & \dfrac{1}{W(\omega_0)} \\ 1 & \cos\omega_1 & \cdots & \cos M\omega_1 & \dfrac{-1}{W(\omega_1)} \\ \vdots & \vdots & \ddots & \vdots & \vdots \\ 1 & \cos\omega_x & \cdots & \cos M\omega_x & \dfrac{(-1)^x}{W(\omega_x)} \end{bmatrix} \cdot \begin{bmatrix} c(0) \\ c(1) \\ \vdots \\ c(M) \\ \delta \end{bmatrix} = \begin{bmatrix} D(\omega_0) \\ D(\omega_1) \\ \vdots \\ D(\omega_x) \end{bmatrix} \qquad (8.38)$$

In Equation 8.37, $\omega_0 - \omega_x$ represent the extremal frequencies and δ is the error. With this expression the filter design problem has been set into a form which can be manipulated by the Remez exchange algorithm.

8.3.2 The Remez Exchange Algorithm

The Remez exchange algorithm is a powerful procedure which uses iteration techniques to solve a variety of minimax problems. (A minimax problem is one in which the best solution is the one that minimizes the maximum error which can occur.) Before initiating the process, a set of discrete frequency points is defined for the passband and stopband of the filter. (Transition bands are excluded.) This dense grid of frequencies is used to represent the continuous frequency spectrum. Extremal frequencies will then be located at particular grid frequencies as determined by the algorithm. The basic steps of the method as it is applied to our filter design problem are shown below.

Remez Exchange Algorithm

I. Make an initial guess as to the location of $x + 1$ extremal frequencies, including an extremal at each band edge.
II. Using the extremal frequencies, estimate the actual frequency response by using the Lagrange interpolation formula.
III. Locate the points in the frequency response where maximums occur and determine the error at those points.
IV. Ignore all new extremals beyond the number initially set in I.
V. If the difference between the maximum and minimum error at the remaining extremal is small enough, continue to VI. Otherwise return to II using the retained extremals.
VI. Estimate the final frequency response and determine the $c(m)$ values from it. Then determine the impulse response coefficients.

Each step in the procedure can be implemented in a variety of ways. These variations can produce differences in the speed of executing the algorithm, but usually little difference in accuracy is noticed. The simplest method of implementing step I is to assign the $x + 1$ extremal frequencies such that they are equally space throughout the bands of interest. Extremals are usually placed at all band edges which are adjacent to transition bands. The initial band and final band may not have extremals located at their terminal edges. The barycentric form of the Lagrange interpolation formula (as described in the mathematical references in Appendix A) is then used to determine the frequency response on the dense grid of frequencies. This method is much more efficient and accurate than the alternative method of finding the $c(m)$ values in Equation 8.37 by matrix inversion. Once the frequency response has been determined, the true extrema can be located and the error at these locations calculated. (Various methods can be used to locate the extrema, usually differing in speed and complexity.) These new frequency points will be used as the new extrema in the next iteration.

It is not unusual to find more extrema in the frequency response than will be needed to characterize the final frequency response. Therefore, some means is necessary to reduce the number of retained frequencies to $x + 1$. Again, there are variations on this procedure, but the general consensus is to retain the extremals which produce the largest error. In step V, we check the difference between the largest and smallest error produced at the retained extremals. By using this value as a progress indicator, we can set some threshold to indicate when the procedure has produced the required level of optimization. If the differences between the minimum and maximum errors have not been reduced enough, the algorithm continues from step II. When the optimization procedure has reached the desired threshold, the extremal frequencies can be used to determine the $c(m)$ values in Equation 8.37 and therefore the impulse response coefficients $h(n)$ from Equation 8.34.

8.3.3 Using the Parks-McClellan Algorithm

The general algorithm has great flexibility in designing any of the four types of FIR filters discussed earlier. The code which is included with this text will design lowpass, highpass, bandpass, and bandstop type 1 filters (with an odd number of symmetrical coefficients). The code is written so that other filter types can be implemented by adding to the program structure. In order to use the general algorithm, we must first convert our filter specifications into those needed by the algorithm. This amounts to converting gain requirements for decibels to absolute error and some redefinition of frequencies.

As in the Kaiser window case, an empirical formulation of the required length of an FIR filter designed using the PM algorithm has been developed as shown in Equation 8.39. Although somewhat extensive in its presentation, it does provide an accurate estimate of the required length.

$$N = \frac{K_1 - K_2 \cdot \Delta f^2}{\Delta f} + 1 \tag{8.39}$$

where

$$\begin{aligned} K_1 = {} & [0.005309 \cdot (\log \delta_p)^2 + 0.07114 \cdot \log \delta_p - 0.4761] \log \delta_s \\ & - [0.00266 \cdot (\log \delta_p)^2 + 0.5941 \cdot \log \delta_p + 0.42781] \end{aligned} \tag{8.40}$$

$$K2 = 0.51244 \cdot (\log \delta_p - \log \delta_s) + 11.012 \tag{8.41}$$

$$\Delta f = (f_{\text{stop}} - f_{\text{pass}}) / f_s \tag{8.42}$$

Example 8.4 Determining Parks-McClellan Coefficients for FIR Filter

Problem: Determine the impulse response coefficients for a bandpass filter using the same specifications as in Example 8.3 (as indicated below), except use the Parks-McClellan algorithm for coefficient determination.

$f_{pass1} = 4$ kHz, $f_{pass2} = 5$ kHz, $f_{stop1} = 2$ kHz, $f_{stop2} = 8$ kHz, $a_{pass} = -0.5$ dB, $a_{stop1} = a_{stop2} = -50$ dB, and $f_{samp} = 20$ kHz

Solution: We will use the FILTER program for this example. The same input parameters as in the previous example are specified, except the approximation type (6) has been selected for PM. Using the specified parameters, the first design attempt resulted in an estimated length of 19, which produced a filter with passband edge gains of –0.54 dB and stopband edge gains of –49.38 dB. These values are certainly very close to the design specifications and might be acceptable in many designs. However, for comparison purposes, the filter was redesigned using a filter length of 21 and produced passband gains of –0.23 dB and stopband gains of –56.70 dB. The resulting coefficients and frequency response curve are shown in Figures 8.15 and 8.16.

```
                    Parks-McClellan FIR Bandpass Filter

                   Filter implementation is:      FIR (digital)
                   Filter approximation is:       Parks-McCl
                   Filter selectivity is:         Bandpass
                   Passband gain          (dB):   -0.50
                   Stopband gain - lower  (dB):   -50.00
                   Stopband gain - upper  (dB):   -50.00
                   Passband freq - lower  (Hz):   4000.00
                   Passband freq - upper  (Hz):   5000.00
                   Stopband freq - lower  (Hz):   2000.00
                   Stopband freq - upper  (Hz):   8000.00
                   Sampling Frequency (Hz):       20000.00

                              Order = 21
                       Overall Gain = 1.00000000e+000

                                  Coefficients
 N    [     N + 0              N + 1              N + 2              N + 3     ]
===   =========================================================================
  0    1.25317859e-002  1.19288691e-003 -3.33806229e-002 -4.32460489e-003
  4    1.22956179e-002  8.30371593e-003  1.02738435e-001 -8.95102562e-003
  8   -2.68102567e-001  4.00023080e-003  3.48855242e-001  4.00023080e-003
 12   -2.68102567e-001 -8.95102562e-003  1.02738435e-001  8.30371593e-003
 16    1.22956179e-002 -4.32460489e-003 -3.33806229e-002  1.19288691e-003
 20    1.25317859e-002
```

Figure 8.15 Coefficient values for Example 8.4.

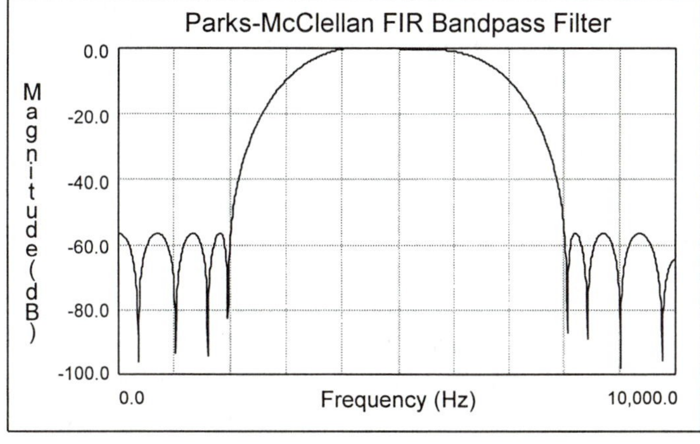

Figure 8.16 Magnitude response for Example 8.4.

We should notice the equal ripple in the stopbands for the PM filter and the fact that it is implemented in one-third fewer coefficients than the Kaiser filter. As a comparison to IIR filters, a sixth-order elliptic or an eighth-order Butterworth filter would be required to satisfy the same specifications, but without linear phase.

8.3.4 Limitations of the Parks-McClellan Algorithm

The Parks-McClellan algorithm is certainly an attractive FIR filter design method, but it does have certain limitations in comparison to window based design methods. The Kaiser window design is basically a one-pass system, although some variation of filter length may be necessary to attain the desired specification (as is also the case for the Parks-McClellan procedure). The computational intensity of the PM method is far greater than for any of the window methods, but computational power is also more readily available than it was 10 years ago. Probably the biggest problem with the PM algorithm is that it does not ***always*** lead to a solution. In some cases, the iteration sequence will not converge, which makes it necessary to place an upper limit on the number of iterations allowed. The algorithm can be modified to allow the frequency grid to be made more dense and to initiate the algorithm again if this happens. Nonetheless, there will be occurrences when the problem statement will need to be redefined in order to attain convergence. For example, if the two stopbands of a bandpass filter are not of approximately equal size, one can be artificially reduced to help the convergence process. Occasionally, the error constraints on the filter must be adjusted to allow more freedom in the optimization process. Even if the process converges, the frequency response of the resulting filter must be checked carefully. Since no requirements are placed on the frequency within the transition bands, strange result can sometimes occur.

8.4 C Code for FIR Filter Design

As we prepare to develop the code for the design of digital FIR filters, we refer to our previous project outline. The only design portion of the outline left for development is item I.B.2, which is shown in expanded form. Item I.B.2 is implemented by the function `Calc_DigFIR_Coefs` shown in Listing 8.1. (All of the functions necessary to implement that section of the project can be found in the \FILTER\F_DESIGN.C module on the software disk.) In this function, we see that all of the items listed under I.B.2 are represented by functions called from `Calc_DigFIR_Coefs`. In addition, the user is given an opportunity to adjust the length of the filter after it has been estimated. This option is necessary since the FIR filter length cannot be calculated exactly. Once the length has been accepted or changed, memory can be allocated for the filter coefficients.

Analog and Digital Filter Design Project

I. Design analog or digital filter.
   A. Get filter specifications from user.
   B. Design filter according to specifications.
      1. Design analog, if designated.
      2. Design digital FIR, if designated.
         a. Estimate length of filter.
         b. If PM design, implement procedure outlined in the previous section.
         c. If window method, then
            (1) Calculate ideal filter coefficients.
            (2) Calculate appropriate window coefficients.
            (3) Multiply ideal and window coefficients.
      3. Design digital IIR, if designated.
   C. Display filter parameters on screen.
   D. Save filter parameters to file.

II. Determine frequency response of filter.

III. Display magnitude and phase response on screen.

```
/*=======================================================
  Calc_DigFIR_Coefs() - calcs the digital FIR coefs
  Prototype:  int Calc_DigFIR_Coefs(Filt_Params *FP);
  Return:     error value
  Arguments:  FP - ptr to struct holding filter params
=======================================================*/
int Calc_DigFIR_Coefs(Filt_Params *FP)
{ char ans;
  int Error;          /*  error value */
  double beta;        /*  parameter for Kaiser window */

  /*  Estimate the length (order) of filter.
      Get beta for Kaiser window, if needed. */
  beta = Estm_Filter_Len(FP);
  /*  See if user wants to adjust estimated length */
  printf("\n Filter length is estimated as %d.",
                                             FP->order);
  ans = Get_YN("\n Do you wish to change it? (Y/N):");
  if(ans == 'Y')
  { FP->order = Get_Int(
```

```
               "\n Please enter new length: ",0,500);}
   /*  Allocate memory for coefficients. */
   FP->acoefs =
         (double *) malloc(FP->order * sizeof(double));
   if(!FP->acoefs) { return ERR_ALLOC;}
   FP->bcoefs =
         (double *) malloc(FP->order * sizeof(double));
   if(!FP->bcoefs) { return ERR_ALLOC;}
   /*  Set overall gain to 1.0 */
   FP->gain = 1.0;
   /*  Calculate the ideal FIR coefficients
       but not for Parks-McClellan. */
   if(FP->approx != '6')
   { Error = Calc_Ideal_FIR_Coefs(FP);
     if(Error) { return 10*Error+1;}
   }
   /*  Determine the approximation method to use.  */
   switch(FP->approx)
   { case '0': Error = Calc_Rect_Win_Coefs(FP);
               if(Error) { return 10*Error+2;}  break;
     case '1': Error = Calc_Bart_Win_Coefs(FP);
               if(Error) { return 10*Error+3;}  break;
     case '2': Error = Calc_Blck_Win_Coefs(FP);
               if(Error) { return 10*Error+4;}  break;
     case '3': Error = Calc_Hamm_Win_Coefs(FP);
               if(Error) { return 10*Error+5;}  break;
     case '4': Error = Calc_Hann_Win_Coefs(FP);
               if(Error) { return 10*Error+6;}  break;
     case '5': Error = Calc_Kais_Win_Coefs(FP,beta);
               if(Error) { return 10*Error+7;}  break;
     case '6': Error = Calc_ParkMccl_Coefs(FP);
               if(Error) { return 10*Error+8;}  break;
     default:  return ERR_FILTER;
   }
   /*  Multiply window and ideal coefs only for
       but not for Parks-McClellan coefficients */
   if(FP->approx != '6')
   { Error = Mult_Win_Ideal_Coefs(FP);
     if(Error) { return 10*Error+9;}
   }
   return ERR_NONE;
}
```

Listing 8.1 Calc_DigFIR_Coefs function.

The length of the FIR filter is estimated by the `Estm_Filter_Len` function. In this function the various parameters required to estimate the length of either a window FIR design or a Parks-McClellan design are calculated. All window designs use the Kaiser estimate, which provides a starting point for the filter designer. In most cases, the Kaiser will be the preferred window design with the other methods used for comparison purposes. After the filter length has been estimated, the calculated value is converted to the next higher odd integer and stored in `FP->order`. We recognize that the variable is actually the filter's length, but this saves us from defining another variable in the `Filt_Params` structure.

If the design method uses the window technique, the ideal filter coefficients are calculated using `Calc_Ideal_FIR_Coefs`. This function implements the appropriate equation for ideal coefficient calculation and stores them in the FP–>bcoefs array. Next, the proper window coefficients are calculated using one of several `Calc_xxxx_Win_Coefs` functions and stored in the FP–>acoefs array. Finally, the ideal and window coefficients are multiplied by the `Multi_Win_Ideal_Coefs` function with the final coefficients stored in the FP–>acoefs. All of these functions are straightforward and will not be listed here. They can be found in the \FILTER\F_DESIGN.C module on the software disk supplied with this text.

If the Parks-McClellan FIR design method has been selected by the user, the `Calc_ParkMccl_Coefs` function shown in Listing 8.2 is called. In this function, one of the first operations is to allocate memory for a structure to hold all of the parameters used in this filter design technique.

```
/*=====================================================
  Calc_ParkMccl_Coefs() - calcs Parks-McClellan coefs
  Prototype: int Calc_ParkMccl_Coefs(Filt_Params *FP);
  Return:    error value
  Arguments: FP - ptr to struct holding filter params
=====================================================*/
int Calc_ParkMccl_Coefs(Filt_Params *FP)
{ int     i,j,k,     /* loop counters */
          Done,      /* process indicator */
          Error;     /* process error number */
  double  max,min,   /* max and min errors */
          test,      /* test value of error */
          Q,Q_old;   /* process quality values */
  PkMc_Params *PM;   /* pointer to Pk-Mc structure */
```

```
if(!FP->acoefs)      /*  Check for valid pointer.  */
{ return ERR_NULL;}
/*  Allocate memory for Parks-McClellan structure */
PM = (PkMc_Params *) calloc(1 , sizeof(PkMc_Params));
if(!PM) { return ERR_ALLOC;}
PM->grid_mult = GRID_MULT/2;
/*  Start Remez outer loop - grid density */
for(k = 0; k < 4 ;k++)
{ PM->grid_mult *= 2;
  /*  Set up all the necessary variables */
  Error = Set_PkMc_Params(FP,PM);
  if(Error) { return 10*Error+1;}
  /*  Start the Remez algorithm inner loop */
  Q_old = 1.0;
  Done = 0;
  /*  Indicate progress */
  printf("\n Working-%d ",k+1);
  for(i = 0; i < 30 ;i++)
  { printf(". ");
    Error = Remez_Interchange(PM);
    if(Error) { return 10*Error+2;}
    /*  Check diff between max and min error */
    max = PM->E[PM->extrml[0]];
    min = max;
    for(j = 1; j < PM->num_ext ;j++)
    { test = PM->E[PM->extrml[j]];
      if(test > max)
      { max = test;}
      if(test < min)
      { min = test;}
    }
    Q = (max - min) / max;
    if(Q < 1E-6)
    { Done = 1;
      break;
    }
    /*  If no significant progress after 15 times,
        try again with increase of grid density */
    if((i > 15) && (Q > Q_old))
    { break;}
    Q_old = Q;
  }
  /*  If done, determine the filter coefs */
  if(Done)
```

```
        { Error = Compute_Coefs(FP,PM);
          if(Error) { return 10*Error+3;}
        }
        /*  Free all memory alloc'd during PkMc routine */
        if(PM->grid_wate)  { free(PM->grid_wate);}
        if(PM->grid_resp)  { free(PM->grid_resp);}
        if(PM->band_pts)   { free(PM->band_pts);}
        if(PM->extrml)     { free(PM->extrml);}
        if(PM->alpha)      { free(PM->alpha);}
        if(PM->beta)       { free(PM->beta);}
        if(PM->grid)       { free(PM->grid);}
        if(PM->x)          { free(PM->x);}
        if(PM->C)          { free(PM->C);}
        if(PM->P)          { free(PM->P);}
        if(PM->E)          { free(PM->E);}

        if(Done)
        { break;}
    }
    if(Done)
    { return ERR_NONE;}
    else
    { return ERR_VALUE;}
}
```

Listing 8.2 Calc_ParkMccl_Coefs function.

The `PkMc_Params` structure is shown in Listing 8.3 and includes all of the necessary variables to control the PM optimization process. The process begins using an outer loop to control the number of grid points used in the frequency response. After setting the number of grid points, all of the necessary PM parameters are initialized by calling the `Set_PkMc_Params` function, which will be discussed later. The process then enters the inner loop in which the Remez interchange is performed by calling the `Remez_Interchange` function. The inner loop is executed for a fixed number of times or until the quality factor `Q` shows no appreciable change. `Q` is defined as the difference between the minimum and maximum errors at the extremal points. If after half of the iterations have been completed, `Q` is still fluctuating instead of steadily decreasing, the process leaves the inner loop. The number of grid frequencies is then increased and the process repeated.

If the outer loop finishes without a satisfactory set of extremals, an error is returned to the calling function. If the process does produce a set of satisfactory extremals, the associated coefficients are computed using the `Compute_Coefs` function. In either case, all memory allocated for the process is freed after each iteration of the outer loop so that new variables can be defined in the `Set_PkMc_Params` function or the process can return to the calling function.

```
typedef struct
{ int     num_pts,        /* total points in grid */
          num_ext,        /* number of extremals */
          num_bands,      /* number of bands */
          filt_type,      /* type of filter */
          grid_mult,      /* grid points / extremal */
          *band_pts,      /* points in each band */
          *extrml;        /* extremal locations */
  double  delta,          /* optimum error */
          *alpha,         /* Lagrange variable */
          *beta,          /* Lagrange variable */
          *C,*x,          /* Lagrange variables */
          *grid,          /* grid points */
          *grid_resp,     /* grid response */
          *grid_wate,     /* grid weight */
          *P,*E;
} PkMc_Params;
```

Listing 8.3 PkMc_Params structure.

The `Set_PkMc_Params` function is quite long and therefore will not be shown in its entirety here. Some of the more routine code will be omitted in Listing 8.4. There are three primary operations performed in this function. First, our filter parameters are converted to a format more appropriate to the PM algorithm, including the determination of the error, the desired response, and weighting factor in each band. Next, the width of and number of grid points in each band are determined. And, finally, extremal frequencies are equally spaced throughout each band. The number of grid points, remember, are increased each time through the outer loop of the calling function.

The `Remez_Interchange` function, which can be found in the \FILTER\FILTER.C module, is another key part of the PM algorithm. It implements the algorithm as described earlier in this section. The code can be viewed on the software disk by using a text editor or word processor.

```
/*=========================================================
  Set_PkMc_Params() - sets up all necessary variables
                      for Parks-McClellan algorithm.
  Prototype: int Set_PkMc_Params(Filt_Params *FP
                                 PkMc_Params *PM);
  Return:    error value
  Arguments: FP - ptr to struct holding filter params
             PM - ptr to struct holding PkMc params
=========================================================*/
int Set_PkMc_Params(Filt_Params *FP,PkMc_Params *PM)
{ /* Declare variables */
  { code omitted }
  /* Determine weighting in stopbands */
  sb_err = pow(10,0.05*FP->astop1);
  pb_err = 1 - pow(10,0.05*FP->apass1);
  W = pb_err / sb_err;
  /*  Set number of bands, array of edge freq, desired
      response and weighting function in each band */
  { code omitted }
  /*  Set filter length and number of distinct coefs,
      assume odd length and symmetric coefs desired,
      set number of extremals and grid points */
  filt_len = FP->order;
  num_cfs = (filt_len + 1) / 2;
  PM->filt_type = SYM_ODD;
  PM->num_ext = num_cfs + 1;
  PM->num_pts = PM->num_ext * PM->grid_mult;
  /*  Allocate memory for various parameters */
  { code omitted }
  /*  Determine the total bandwidth of
      passbands and stopbands */
  BW_total = 0;
  for(b = 0; b < PM->num_bands ;b++)
  { band_wid[b] = (edge[2*b + 1] - edge[2*b]);
    BW_total += band_wid[b];
  }
  /*  Determine grid pts and extrmls in each band */

  grid_dens = (double)
             (PM->num_pts-PM->num_bands) / BW_total;
```

```
  extr_dens = (double)
            (PM->num_ext-PM->num_bands) / BW_total;
  band_ext[PM->num_bands-1] = PM->num_ext;
  PM->band_pts[PM->num_bands-1] = PM->num_pts;
  for(b = 0; b < PM->num_bands-1 ;b++)
  { PM->band_pts[b] =(int)(grid_dens*band_wid[b]+1.5);
    PM->band_pts[PM->num_bands-1] -= PM->band_pts[b];
    band_ext[b] = (int)(extr_dens * band_wid[b]+1.5);
    band_ext[PM->num_bands-1] -= band_ext[b];
  }
  /*  Set up dense grid, set response and weight */
  j = 0;
  PM->grid[0] = 0;
  for(b = 0; b < PM->num_bands ;b++)
  { freq_spc =band_wid[b]/(double)(PM->band_pts[b]-1);
    PM->grid[j] = edge[2 * b];
    PM->grid_resp[j] = resp[b];
    PM->grid_wate[j++] = wate[b];
    for(i = 1; i < PM->band_pts[b] ;i++)
    { PM->grid[j] = PM->grid[j-1] + freq_spc;
      PM->grid_resp[j] = resp[b];
      PM->grid_wate[j++] = wate[b];
    }
  }
/*  Equally space initial extrml freqs along grid in
      each band. Store grid index where max's occur */
  e = 0;
  index = 0;
  for(b = 0; b < PM->num_bands ;b++)
  { pt_dens = (double)PM->band_pts[b] /
                               (double)band_ext[b];
    PM->extrml[e++] = index;
    for(i = 1; i < band_ext[b] ;i++)
    { PM->extrml[e++] =
          (int)((double)i * pt_dens + 0.5) + index;
    }
    index += PM->band_pts[b];
  }
  return ERR_NONE;
}
```

Listing 8.4 Set_PkMc_Params function.

8.5 C Code for FIR Frequency Response Calculation

Now that the FIR filter coefficients have been determined, our last task in developing this complete filter design project is to determine the frequency response of the filter. As determined in Chapter 6, the frequency response of a digital filter can be determined from the transfer function as shown in Equation 8.43.

$$H(e^{j\Omega}) = H(z)\big|_{z=e^{j\Omega}} \tag{8.43}$$

In the case of a causal FIR filter, the transfer function can be described as

$$H(z) = \sum_{k=0}^{N-1} h(k) \cdot z^{-k} \tag{8.44}$$

which then leads to the following description for the frequency response.

$$H(e^{j\Omega}) = \sum_{k=0}^{N-1} h(k) \cdot e^{-jk\Omega} \tag{8.45}$$

This expression can also be expressed as a sum of the real and imaginary portions of the exponential.

$$H(e^{j\Omega}) = \sum_{k=0}^{N-1} h(k) \cdot \cos(k\Omega) + j\sum_{k=0}^{N-1} h(k) \cdot \sin(k\Omega) \tag{8.46}$$

This expression can be simplified in a variety of ways based on the symmetry of the coefficients. However, to keep the response calculation as general as possible, the `Calc_DigFIR_Resp` function will implement Equation 8.46 directly as shown in Listing 8.5.

The magnitude variable is initialized to the value of the gain constant, and the angle variable is set to zero. Then the response at each frequency is determined by evaluating the effect of each filter coefficient. The angle is converted to degrees, and after all calculations have been made, the magnitude is converted to decibels if the user requested that format.

```
/*=========================================================
  Calc_DigFIR_Resp() - calcs response for FIR filters
  Prototype:  int Calc_DigFIR_Resp(Filt_Params *FP,
                                    Resp_Params *RP);
  Return:     error value
  Arguments:  FP - ptr to struct holding filter params
              RP - ptr to struct holding respon params
              fsamp - sampling frequency (Hz)
=========================================================*/
int Calc_DigFIR_Resp(Filt_Params *FP,Resp_Params *RP)
{ int     f,i;           /*  loop counters */
  double  rad2deg,       /*  rad to deg conversion */
          omega,i_omega,/*  radian freq and incrmnt */
          mag,           /*  magnitude of freq resp */
          rea,img;       /*  real and imag part */

  rad2deg = 180.0 / PI; /* set rad2deg */
  /*  Loop through each of the frequencies */
  for(f = 0 ;f < RP->tot_pts; f++)
  { /* Initialize magna and angle */
    RP->magna[f] = FP->gain;
    RP->angle[f] = 0.0;
    /* Pre calc adjusted omega, rea and img */
    omega = PI2 * RP->freq[f] / FP->fsamp;
    rea = 0.0; img = 0.0;
    /* Loop through all the coefs */
    for(i = 0 ;i < FP->order; i++)
    { i_omega = i * omega;
      rea += FP->acoefs[i] * cos(i_omega);
      img += FP->acoefs[i] * sin(i_omega);
    }
    /* Calc final result and conv to degrees */
    mag = sqrt(rea*rea + img*img);
    RP->magna[f] *= mag;
    /*  Guard against atan(0,0) */
    if(mag > 0)
    { RP->angle[f] += atan2(img,rea);}
    RP->angle[f] *= rad2deg;
  }
  /*  Convert magnitude response to dB if indicated */
  if(RP->mag_axis == LOG)
  { for(f = 0 ;f < RP->tot_pts; f++)
    { /* Handle very small numbers */
      if(RP->magna[f] < ZERO)
```

```
            { RP->magna[f] = ZERO;}
            RP->magna[f] = 20 * log10(RP->magna[f]);
          }
        }
        return ERR_NONE;
      }
```

Listing 8.5 Calc_DigFIR_Resp function.

8.6 Conclusion

By completing this chapter, we have completed the material on digital filter design. We have investigated two of the most popular methods of FIR filter design: the Fourier series method using window functions and the Parks-McClellan optimization method. We can also determine and display the magnitude and phase response of the FIR filters we design. In the next chapter we will investigate the implementation of both the FIR and IIR digital filters we have designed.

Chapter 9

Digital Filter Implementation Using C

In the previous three chapters we have discussed the nature of digital filter design. We are now ready to discuss the implementation of these digital filters. We will begin this chapter with a discussion of several important issues in digital filter selection and implementation. These issues include the differences between real-time and nonreal-time implementation, as well as the effects of finite precision representation of input signals and filter coefficients. Then we will discuss the C code for implementing IIR and FIR filters. Efficient algorithms will be developed to increase the speed of execution. Each filter type will use a different technique appropriate to the specific filter's representation. And finally, we will conclude with a discussion of the format for a couple of popular sound file formats on the PC. We will consider how we can use sound files to investigate the characteristics of the filters we have designed.

9.1 Digital Filter Implementation Issues

The first decision which must be made when designing a system which will use a digital filter is whether an IIR or FIR filter should be used. Some of the advantages and disadvantages of each type have been discussed in the previous two chapters, so we will summarize those points here. First, and foremost, the correct filter type must be determined by the requirements of the application. IIR (recursive) filters have the advantages of providing higher selectivity for a particular order and a closed form design technique which doesn't require iteration. The design technique also provides for the rather precise solution to the specifications of gain and edge frequencies. However, IIR filters also have the disadvantages of nonlinear phase characteristics and possible instability due to poor implementation. FIR (nonrecursive) filters, on the other hand, can provide a linear phase response (constant group delay) which is important for data transmission and high-quality audio systems. Also, they are always stable because they are implemented using an all-zero transfer function. Since no poles can fall outside the unit circle, the filter will always be stable. But because of this, the order of the filter is much higher than the IIR filter which has a comparable magnitude response. This higher order leads to longer processing times and larger memory requirements. In addition, FIR filters must be designed using an iterative method since the required filter length to satisfy a given filter specification can only be estimated.

Therefore, the filter designer must weight the requirements placed on the digital filter. If great importance is placed on magnitude response with much less importance on phase response, then an IIR filter would seem the better choice. If phase response if far more important than magnitude response, then an FIR filter is in order. If both magnitude and phase response seem to be of equal concern, then the processing time constraints and memory requirements must be considered. The FIR filter, even when designed using the Parks-McClellan method, will require more processing time and more memory to implement, but always will be stable. If all else fails, both an FIR and IIR filter (with some phase correction) can be designed to meet the specifications and they both can be tested to evaluate the results.

Once the choice of filter type has been made, there are still a number of decisions which must be made. For example, is the system to operate in real-time or can it be a nonreal-time system? A real-time system is one in which input samples are provided to the digital filter and must be processed to provide an output sample before the next input sample arrives. Obviously, this puts a very precise time constraint on the amount of processing available to the system. The higher the sampling frequency, the less time is available for processing. On the other hand, some systems are afforded the luxury of being able to operate in nonreal time. For example, signals can be recorded on tape or other media and processed at a later time. In this type of system, extensive processing can take place, because there is no fixed time interval which marks the end of the processing time. In nonreal-time applications, high-precision floating point representation for the coefficients and signal can usually be used since speed is not the critical factor. However, the majority of digital filter applications will be real-time applications. In these applications there will inevitably be a battle to obtain the highest accuracy, the fastest speed, and the lowest-cost system. One of the first decisions which must be made is whether the system will be a fixed point or a floating point system. This deals not only with the input and output signal streams, but also the representation of the coefficients and intermediate results within the processing unit.

9.1.1 Input and Output Signal Representation

At this point in time, most digital data to be processed by digital filters is in the form of fixed point data. The reason for this is that even if the original form of the data is in continuous form (best represented by floating point notation), it is converted to digital form by an analog-to-digital (A/D) converter which provides a fixed point representation. Although most input and output digital signals use fixed point representation, there are several ways in which the same signals can be interpreted. Most people are familiar with the base 10 system that humans use, but the digital computers of the world use a binary or base 2 system. In that system, each digit represents a multiplier of a power of 2 just as each digit in the decimal or base 10 system represents a multiplier of a power of 10. Let's start with an 8-bit binary number shown below.

$$10011000_2 = 1 \cdot 2^7 + 1 \cdot 2^4 + 1 \cdot 2^3 = 152_{10} \tag{9.1}$$

This binary number is equivalent to 152_{10} only when we assume that the number is unsigned (represents only positive numbers). If we had considered the binary number as signed, the leftmost 1 would have indicated a negative sign, and the value would have been interpreted differently. Using two's complement arithmetic, we can determine the value of the number in Equation 9.1 by first negating every digit in the number and then adding one to the result. That value is then considered a negative value as shown in Equation 9.2.

$$10011000_2 = -(01100111_2 + 1) = -01101000_2 = -104_{10} \tag{9.2}$$

We can even interpret the original binary number in a different way if we assume that the placement of the binary point (equivalent to the decimal point) is located at a position other than to the right of the right most digit. For example, if we place the binary point in the middle of the eight binary digits, the number takes on another value entirely. (In the representation shown below, it is considered an unsigned number, but it could just as well be considered a signed number as described before.)

$$1001.1000_2 = 1 \cdot 2^3 + 1 \cdot 2^0 + 1 \cdot 2^{-1} = 9.5_{10} \tag{9.3}$$

There are advantages and disadvantages to each of the binary representations. The use of signed numbers is required in most digital filters, and the two's complement representation provides easier methods for addition and subtraction, but multiplication requires special consideration. The signed fractional arithmetic illustrated in Equation 9.3 provides great efficiency in multiplication even though addition and subtraction do require some special steps. By using fractional notation, as most commercial digital signal processing (DSP) chips do, we can guarantee that the result of a multiplication will still be less than one and therefore cause no overflow. However, the addition of two fractional numbers can provide a result larger than one and therefore must be handled carefully. Techniques for dealing with potential errors in calculations are considered further in Section 9.1.3.

Before we leave the area of input and output data representation completely, we must consider the effect of the number of bits per sample on the signal-to-noise ratio (SNR) of a digital system. It is shown in a number of the digital filter design references listed in Appendix A that the SNR for a digital signal processing system will be directly proportional to the number of quantization bits used. To be specific, the equation is

$$\mathrm{SNR}_{\mathrm{dB}} = 6.02 \cdot B + 20 \cdot \log(\sigma_x \,/\, A_{\max}) + 10.8 \tag{9.4}$$

In Equation 9.4, B represents the number of bits used to represent the magnitude of the digitized signal, while σ_x and $A_{\max}$ represent the input signal's standard deviation and the quantizer's maximum input signal. Usually, the input signal's amplitude is adjusted such that $\sigma_x/A_{\max}$ is approximately 1/4. If this condition is met, and we assume that the input signal has a Gaussian distribution (which is a valid assumption for many signals), then the quantizer's limits will be exceeded only 0.064% of the time. (If a smaller number is chosen for the ratio, the dynamic range of the system would be sacrificed. If a larger number is chosen, the frequency of overflow would increase.) Using the value of 1/4 gives

$$\mathrm{SNR}_{\mathrm{dB}} = 6.02 \cdot B - 1.2 \tag{9.5}$$

An important realization drawn from this expression is that the SNR can be improved by 6 dB for every bit added to the quantized representation. For example, if a particular filtering application requires a signal-to-noise ratio of 80 dB, then a D/A converter with at least 14 bits of magnitude quantization must be available.

9.1.2 Coefficient Representation

When it comes to the internal processing of data for a digital filter system, there is more of an even mix between fixed point and floating point systems. The market has a variety of DSP microprocessor chips on the market from a number of manufacturers. These DSP systems include a wide selection of fixed point and floating point systems. The fixed point systems generally provide higher processing speed at lower cost than do the floating point systems. However, the floating point systems provide accuracy which many fixed point systems can not achieve. The representation for the coefficients does not have to be the same as the representation selected for the input signal. Even if both are fixed point numbers, the coefficient representation can use a higher precision representation (more bits). It is up to the digital filter designer to determine the system characteristics such that sufficient accuracy is achieved with adequate processing speed at the lowest possible cost.

The representation of the filter coefficients within the DSP system is of prime concern. Enough accuracy is required in the representation of the filter coefficients (which determine the pole and zero locations) to guarantee that the specifications of the filter are met by the implementation. If the accuracy of the coefficients is compromised, the response of the filter may be severely distorted, and in some cases, the stability of IIR filters can be jeopardized. FIR filters will always be stable because they are represented by transfer functions with all zeros. However, their frequency response can still be affected by the lack of accuracy of their coefficients. As an example, we can consider the case of the Parks-McClellan filter designed in Example 8.4. The coefficients for that example have been truncated to signed 16-bit, 12-bit, and 8-bit numbers as shown in Table 9.1.

Table 9.1 Comparison of original and truncated coefficients.

Coefs	Original	16-bit	12-bit	8-bit
$h(10)$	0.348855	0.348855	0.348855	0.348855
$h(9),h(11)$	0.004000	0.004003	0.003920	0.002747
$h(8),h(12)$	-0.268103	-0.268101	-0.268075	-0.269195
$h(7),h(13)$	-0.008951	-0.008954	-0.009032	-0.008241
$h(6),h(14)$	0.102738	0.102739	0.102765	0.101635
$h(5),h(15)$	0.008304	0.008304	0.008351	0.008241
$h(4),h(16)$	0.012296	0.012297	0.012270	0.010988
$h(3),h(17)$	-0.004325	-0.004322	-0.004261	-0.005494
$h(2),h(18)$	-0.033381	-0.033377	-0.033403	-0.032963
$h(1),h(19)$	0.001193	0.001192	0.001193	0.000000
$h(0),h(20)$	0.012532	0.012531	0.012611	0.013734

Equation 9.6 shows the procedure for determining the truncated values. The calculation within the Round function actually indicates the procedure of converting the floating point coefficient to a signed fraction representation for use in fixed point processors. The multiplication outside of the Round function converts the signed fraction back to a truncated decimal.

$$h_{\text{trunc}}(n) = \text{Round}\left[\frac{h(n)\cdot(2^{B-1}-1)}{h(0)}\right]\cdot\frac{h(0)}{(2^{B-1}-1)} \tag{9.6}$$

As we would expect, as the number of bits is reduced, the error in the coefficients increases. The frequency response generated from the 16-bit coefficients was virtually identical to the original response. However, the responses due to the 12-bit and 8-bit coefficients were noticeably degraded as shown in Figure 9.1. In that figure, the passband response was virtually unchanged in all cases, but the stopband characteristics did not match those of the original coefficients (shown as –56.7 dB). The 12-bit coefficients did produce a response which satisfied the original design specifications of –50 dB. The 8-bit coefficients, however, caused the frequency response to degenerate completely, providing as little as 32 dB of attenuation. The reason for this degradation, of course, is that the truncation has moved the pole and zero locations from their original positions.

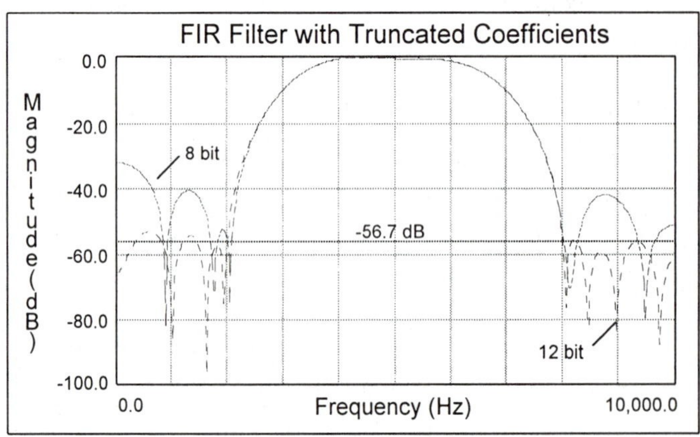

Figure 9.1 FIR filter frequency response using truncated coefficients.

9.1.3 Retaining Accuracy and Stability

There is no effective way to directly relate the accuracy of the coefficients to the degree of degradation of the frequency response. At this time, trial and error techniques are all that can be offered when determining the necessary accuracy of coefficients. However, there are some helpful practices which can be observed when dealing with the implementation of these coefficients to reduce the effects of truncation. The following suggestions relate to IIR filters unless stated otherwise since they have some unique problems due to their recursive nature. The fact that IIR filters are implemented using feedback leads to special problems which the FIR filter does not experience.

Probably the most important implementation rule when dealing with IIR filters is that it is much better to implement the filter as a cascade of quadratic factors (as we have done) than to combine the transfer function into a quotient of high-order polynomials. (Even some experimentation has been done with this idea for FIR filters.) This technique provides better control on the stability of the filter. By quantizing the coefficients which represent the filter, we are actually specifying a fixed number of positions within the unit circle where the poles can be located. The fewer bits used for quantization, the fewer the positions for the poles. In addition, these pole locations are not uniformly distributed within the unit circle. However, by choosing different topologies for the implementation (such as a coupled form of quadratic), more uniformly distributed pole location can be achieved. Also, it has been shown that those poles which either lie close to the unit circle or to each other are the most critical to represent properly, and therefore may need some special implementation method. Another technique which can be used to reduce errors caused by computations using finite accumulators is to pair poles and zeros which are located near each other in the same quadratic factor to reduce large fluctuations. In addition, sections with poles closest to the unit circle can be moved to the end of the evaluation process to help eliminate overflows. The digital filter design references in Appendix A provide further information on other structures than the quadratic form, including those for representing poles located extremely close to the unit circle. Antoniou's, Oppenheim/Shafer's (1975) and Proakis/Manolakis's texts provide valuable and detailed material.

Several other points can be made when discussing the coefficients and internal processing of the filter calculations. In both FIR and IIR filter implementation, the final output values are stored in a register which gradually accumulates the value of the final output. This accumulator should normally be allocated as many bits of representation as possible because it controls the ultimate accuracy of the processor. Overflow and underflow are potential problems for the accumulator since it is hard to predict the exact nature of incoming signals. Most present dedicated DSP chips provide some form of scaling which can be applied to the input so that temporary large variations can be accommodated without incurring a great deal of error. This scaling bit can effectively be used as a temporary additional bit of accuracy for accumulated values to prevent overflow. However, if overflow is inevitable, it is better to have a processor which will simply saturate at its maximum level than to allow an overflow which can be interpreted as a swing from positive to negative value.

In recursive systems which use finite precision representations, a troublesome problem called limit cycle oscillation can occur. There are actually two types of limit cycles: overflow and quantization. Overflow limit cycles (also called large-signal limit cycles) are due to the unmanaged overflow of the accumulator during processing. Most overflow limit cycles can be effectively eliminated by the proper use of signal scaling at the input of the system and saturation arithmetic in the accumulator. Many DSP processors include some scaling feature within the processor unit which allows the input signal to be reduced in size. However, a reduction in the size of the input signal also reduces the SNR of the system although this is usually better than the distortion produced by an overflow. Another useful feature on many DSP systems today is the use of saturation arithmetic within the processing unit. This feature will saturate the value to its positive or negative limit rather than overflow the accumulator. Again, this will result in distortion, but usually less than the overflow would produce. Of course, another way to help control overflow limit cycles is to increase the size of the accumulator, if the processing system allows the accumulator to be larger than the standard coefficient storage size.

The quantization limit cycles (also called small-signal limit cycles) generally result from the handling of quantization within the system and are noticeable when the output should be constant or zero. This problem is the result of the input signal changes being less than the quantization level. Two methods have been developed to combat this type of limit cycle. The first, which may not be a practical alternative, is to increase the number of bits assigned to the representation of the signal values in order to reduce the quantization error. By reducing the quantization error, the limit cycles can either be eliminated altogether, or reduced to a nonobjectionable level. The second method suggests that the products produced by finite precision multiplication be truncated, rather than rounded, as they are accumulated. Other, more detailed, analysis of limit cycles is included in the references.

9.2 C Code for IIR Filter Implementation

When discussing the implementation of IIR filters, we will assume that the filter is described by a set of quadratic coefficients of the form determined in Chapter 7. As we have seen in the previous section, a cascaded sequence of quadratic structures is the recommended method of implementation. The basic quadratic building block is shown in the system diagram of Figure 9.2.

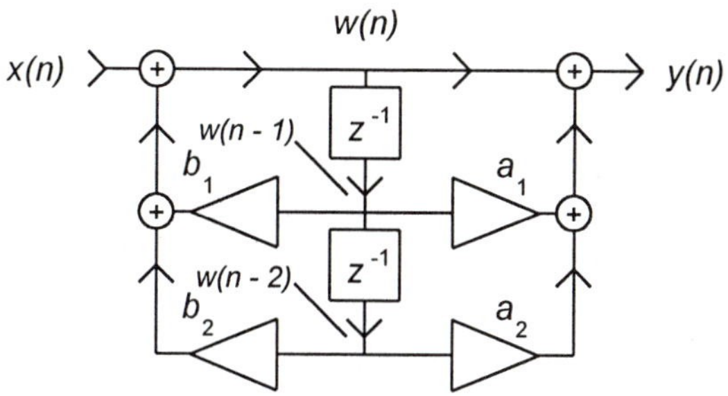

Figure 9.2 System diagram for a single quadratic factor.

We can generate the transfer function for this section by determining the expressions for the intermediate signal $w(n)$ and the output signal $y(n)$.

$$w(n) = x(n) + b_1 \cdot w(n-1) + b_2 \cdot w(n-2) \tag{9.7}$$

$$y(n) = w(n) + a_1 \cdot w(n-1) + a_2 \cdot w(n-2) \tag{9.8}$$

These equations can be z-transformed to give

$$W(z) = X(z) + b_1 \cdot z^{-1} \cdot W(z) + b_2 \cdot z^{-2} \cdot W(z) \tag{9.9}$$

$$Y(z) = W(z) + a_1 \cdot z^{-1} \cdot W(z) + a_2 \cdot z^{-2} \cdot W(z) \tag{9.10}$$

Equations 9.9 and 9.10 can be rewritten and combined to determine the transfer function for this section of the filter as shown in Equation 9.11. This formulation matches the quadratic terms we developed for IIR filters. We will be able to match equivalent terms if we recognize two characteristics. First, the a_0 coefficient is always one, and second, the b coefficients in the system diagram will be the negative of their value in the transfer function equation.

$$H(z) = \frac{Y(z)}{X(z)} = \frac{(1 + a_1 \cdot z^{-1} + a_2 \cdot z^{-2})}{(1 - b_1 \cdot z^{-1} - b_2 \cdot z^{-2})} \tag{9.11}$$

As an example, consider the fourth-order transfer function for a Chebyshev highpass filter shown in Equation 9.12. This filter can be implemented by using the system diagram shown in Figure 9.3. As we see in the diagram, the input signal is first multiplied by the gain constant and is then processed through two quadratic factors. The multiplication by the gain constant could occur at the end of the process or be distributed throughout the diagram as well. Notice the sign difference on the b coefficients between the transfer function of Equation 9.12 and the system diagram of Figure 9.3. (Of course, the coefficients will be represented with higher precision than indicated here.)

$$H_{C4}(z) = \frac{0.201 \cdot (1 - 2 \cdot z^{-1} + 1 \cdot z^{-2}) \cdot (1 - 2 \cdot z^{-1} + 1 \cdot z^{-2})}{(1 - 1.047 \cdot z^{-1} + 0.795 \cdot z^{-2}) \cdot (1 - 0.0448 \cdot z^{-1} + 0.222 \cdot z^{-2})} \quad (9.12)$$

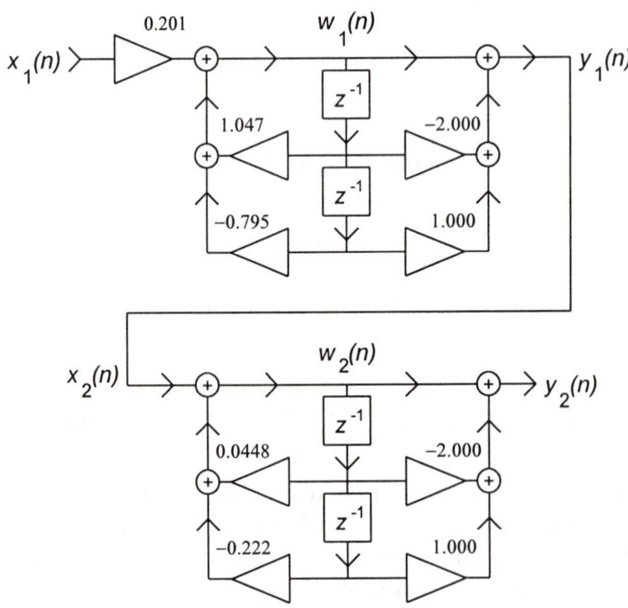

Figure 9.3 System diagram for fourth-order IIR filter.

After determining the system diagram for a filter, we can use the diagram as a guide to implementing the filter. For every quadratic factor, we can calculate an intermediate signal $w(n)$ and an output signal $y(n)$. Besides $x(n)$, $w(n)$, and $y(n)$, every quadratic section also requires the values $w(n-1)$ and $w(n-2)$ (as shown in Figure 9.2) to be retained. We can rewrite Equations 9.7 and 9.8 as

$$w = x + b_1 \cdot m_1 + b_2 \cdot m_2 \tag{9.13}$$

$$y = w + a_1 \cdot m_1 + a_2 \cdot m_2 \tag{9.14}$$

where we have defined

$$m_1 = w(n-1) \tag{9.15}$$

$$m_2 = w(n-2) \tag{9.16}$$

The names m_1 and m_2 are picked to reflect the fact that these values are the memory states of the quadratic factor. Each quadratic structure must keep track of the previous values which have been present in the structure.

The IIR filtering process can be implemented by first multiplying the input signal by the gain and then implementing Equations 9.13 – 9.16 for each of the quadratics. Then, before progressing to the evaluation of the next quadratic factor, the values of m_1 and m_2 are updated. A section of code which will implement this process is shown in Listing 9.1. In the listing, it is assumed that there are `m1` and `m2` arrays with sizes equal to the number of quadratics, and that the `a` and `b` coefficients for the filter are stored in the usual manner. (That is, the three coefficients a_0, a_1, and a_2 for the first quadratic are followed by the three coefficients for the second quadratic and so on. The same organization is assumed for the b coefficients.) Notice that the a_0 and b_0 coefficients are not used since they are assumed to be one. We have also substituted the output variable `o` for the input variable x to allow the output value to accumulate through several quadratic sections.

```
o = x * gain;
for(j = 0; j < numb_quads ;j++)
{ jj = j*3;
  w = o - m1[j] * b[jj+1] - m2[j] * b[jj+2];
  o = w + m1[j] * a[jj+1] + m2[j] * a[jj+2];
  m2[j] = m1[j];
  m1[j] = w;
}
```

Listing 9.1 Segment of code for IIR filter implementation.

Although the algorithm in Listing 9.1 will make the correct computations, it could be slow because of all of the index calculations which must be made. (Actually, compilers will differ in terms of how much optimization can be made with speed critical code such as we are discussing.) The speed of this loop can usually be increased by using pointers to the coefficients and memory values. In order to take advantage of this pointer efficiency we must store the filter coefficients in an orderly manner which will allow them to be accessed sequentially in the exact order that they are needed. We can define a new array `C` which will store all of the needed coefficients as well as the filter gain constant. The first element in the array will be the gain constant followed by the coefficients b_1, b_2, a_1, and a_2 for each quadratic factor. The structure of the `C` array then has the following form.

$C[0] = \text{gain}$	$C[5] = b_1$ (quad 2)
$C[1] = b_1$ (quad 1)	$C[6] = b_2$ (quad 2)
$C[2] = b_2$ (quad 1)	$C[7] = a_1$ (quad 2)
$C[3] = a_1$ (quad 1)	$C[8] = a_2$ (quad 2)
$C[4] = a_2$ (quad 1)	• • •

In addition, an `M` array must be created to store the memory states of the IIR filter. The size of the array is equal to twice the number of quadratic factors with the m_1 states stored in the first half of the array and the m_2 states in the last half. Initially, all of the memory states are set to zero (unless we know some predefined state exists for the filter). Thus the `M` array has the following structure where N is the number of quadratics.

$M[0] = m_1$ (quad 1)	$M[N] = m_2$ (quad 1)
$M[1] = m_1$ (quad 2)	$M[N+1] = m_2$ (quad 2)
$M[2] = m_1$ (quad 3)	$M[N+2] = m_2$ (quad 3)
• • •	• • •

Listing 9.2 shows the `Dig_IIR_Filter` function used in the DIGITAL program discussed later in this chapter. (All of the functions discussed in this chapter can be found in the \DIGITAL\DIGITAL.C module on the software disk which accompanies this text.) The function takes as arguments pointers to arrays of input values (`X`), output values (`Y`), memory states (`M`), and coefficient values (`C`), as well as the number of quadratic factors and number of values in the input and output arrays. Notice that since we will be progressing through the `X`, `Y`, `C`, and `M` arrays, we have defined indexing

pointers x, y, c, and m which can take on changing values. (*Never* use the array name itself as an index or the address of the array will be lost.) The notation may look a little foreign so let's take a look at the code on a line by line basis. (The line numbers shown are to help with this discussion and are not part of the normal code.) The process of using pointers is very efficient for computational purposes because it fits the nature of the computer. However, it can be a bit confusing to follow. Therefore, in order to provide a more descriptive analysis of the IIR filter code, the code has been annotated with the status of primary arrays and variables. For this illustration, it is assumed that there are two quadratic factors and sample values have been entered for the C and M arrays. At each step of the process, the particular values to which c, m1, or m2 are pointing within the arrays are shown in bold. Only the first time through the inner loop is illustrated, but we can see from this how the pointers progress through the arrays, and how the memory state array is changed.

In lines 1 and 2 of Listing 9.2, the addresses of the input and output arrays are transferred to temporary variables x and y which can change without affecting the addresses stored in X and Y. Then line 4 begins a for loop to process every value in the input array. Lines 6 – 8 set up pointers to the memory states and coefficient arrays. The m1 pointer is set to start at the beginning of the M array where the m_1 states are stored, while the m2 pointer is set midway through the M array since the m_2 states are stored in the second half of the array. The initial value of the output, indicated by o, is calculated by multiplying the input value by the first value in the coefficient array, which is the gain constant.

```
/*=======================================================
  Dig_IIR_Filter() - filters input array using IIR
                coefs and mem values to generate output
  Prototype:  void Dig_IIR_Filter(int *X,int *Y,
              double *M,double *C,int numb_quads,int N);
  Return:       error value.
  Arguments:  X - ptr to input array
              Y - ptr to output array
              M - ptr to memory array
              C - ptr to coefs array
              numb_quads - number of quadratics
              N - number of values in array
=======================================================*/
void Dig_IIR_Filter(int *X,int *Y,double *M,double *C,
                                  int numb_quads,int N)
```

```
{ int      *x,*y,      /*  ptrs to in/out arrays */
           i,j;        /*  loop counters */
  double   *c,*m1,*m2, /*  ptrs to coef/memory arrays*/
           w,o;        /*  intermed & output values */

  /*  Make copies of input and output pointers */
    x = X;
    y = Y;
```

*x: **3**,4,5,6,7,...*

*y: **0**,0,0,0,0,...*

```
    /*  Start loop for number of data values */
    for(i = 0; i < N ;i++)
    { /*  Make copies of pointers and start calcs */
      m1 = M;
      m2 = M + numb_quads;
      c = C;
```

*C: **0.20**,1.0,-0.80,-2.0,-1.0,0.05,-0.20,-2.0,-1.0*

*M: **1.0**,3.0,**2.0**,4.0*

```
      o = *x++ * *c++;
```

*C: 0.20,**1.0**,-0.80,-2.0,-1.0,0.05,-0.20,-2.0,-1.0*

*x: 3,**4**,5,6,7,...*

o: 0.60

```
      /*  Start loop for number of quad factors */
      for(j = 0; j < numb_quads ;j++)
      { w = o - *m1 * *c++;
```

*C: 0.20,1.0,**-0.80**,-2.0,-1.0,0.05,-0.20,-2.0,-1.0*

*M: **1.0**,3.0,**2.0**,4.0*

w: -0.40

```
        w -= *m2 * *c++;
```

*C: 0.20,1.0,-0.80,**-2.0**,-1.0,0.05,-0.20,-2.0,-1.0*

*M: **1.0**,3.0,**2.0**,4.0*

w: 1.20

```
        o = w + *m1 * *c++;
```

*C: 0.20,1.0,-0.80,-2.0,**-1.0**,0.05,-0.20,-2.0,-1.0*

*M: **1.0**,3.0,**2.0**,4.0*

o: -0.80

```
        o += *m2 * *c++;
```

*C: 0.20,1.0,-0.80,-2.0,-1.0,**0.05**,-0.20,-2.0,-1.0*

*M: **1.0**,3.0,**2.0**,4.0*

o: -2.8

```
        *m2++ = *m1;
```

*M: **1.0**,3.0,1.0,**4.0***

```
        *m1++ = w;
```

```
                M: 1.2,3.0,1.0,4.0
18     }
19     /*  Convert output to int and store */
20     *y++ = (int)ceil(o-0.5);
                y: -3,0,0,0,0,...
21   }
}
```

Listing 9.2 Dig_IIR_Filter function.

Notice that the method used to access the gain constant is by using the `*c` notation, which can be read as "the value at which *c* is pointing." Since `c` has just been initialized to the start of the `C` array, and since the first element in `C` is the gain constant, then `c` is pointing at the gain constant. The "++" notation after `c` instructs the computer to increment the value of `c` by one *after* the equation has been evaluated. (This is referred to as post-incrementing as opposed to `++c`, which is called preincrementing.) We are performing pointer arithmetic at this point which is a special type of arithmetic. Remember that the variable `c` (and `C`) contains a number which is an address of where coefficients are stored in memory. When we increment a pointer we are incrementing an address and therefore must do so in a meaningful way. In this case, the coefficients are stored as `double's`, which take up 8 bytes of memory on a PC. It would be meaningless to increment the address of an array containing `double's` by 1 byte, or 2, or anything but 8 bytes. Therefore, when we tell the machine to increment `c`, it increments the number stored in `c` by 8. (This is what is meant by special arithmetic.) The compiler keeps track of how to increment pointer variables based on how the variables are initially declared. For example, on a PC, if the variables are declared as `double's`, the increment is 8; if they are declared as `int`'s, the increment is 2.

Now, returning to our discussion of Listing 9.2, we next enter a `for` loop (line 11) used to calculate the effects of each quadratic factor on the ultimate output value. Lines 12 – 13 compute the value of `w` and increment the pointer in the `C` array as each value is used. Lines 14 – 15 then make the calculations representing the right half of Figure 9.2 to determine the output value. The pointer `c` is now pointing to the `b` coefficients for the next quadratic section. Lines 16 – 17 update the values of the memory states and increment `m1` and `m2` to point to the memory states of the next quadratic section. The final step in the determination of the output value is the conversion of the floating point value of `o` to the fixed point value of `*y`. The method used in converting the floating point value to an integer value insures that both positive and negative values will be rounded correctly. The pointer `y` is then incremented and the process repeats for the next value in the input array. We should note that by the end of the outer `for` loop, the `c,m1`, and `m2` pointers will be pointing to values which don't belong to them. That's all right as long as we don't try to access them. The next time an input value is filtered, lines 6 – 8 will reset the pointers to the correct positions.

One advantage of the algorithm we have just developed is that it can be used effectively for either real-time or nonreal-time applications. The value x can be taken from either an input port of a DSP system, or from an array of values to be processed (as illustrated). Likewise, the value `y` can be fed to an output port of a DSP system or placed in an output array (as we have done). We will see a program for a complete filtering system in Section 9.4.

9.3 C Code for FIR Filter Implementation

An FIR filter has no feedback and thus its system diagram can be displayed as shown in Figure 9.4. This configuration represents a convolution involving N coefficients as described by Equation 9.17.

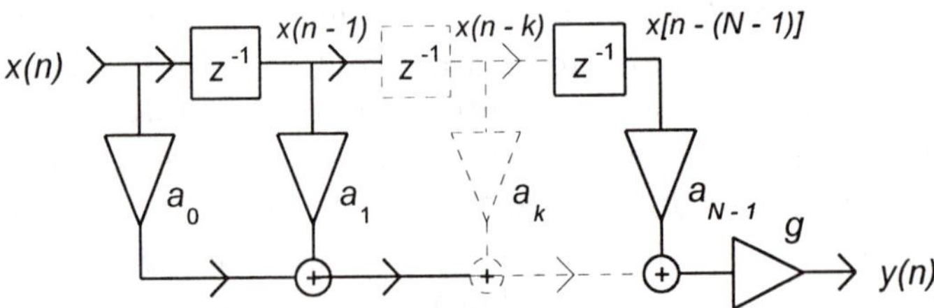

Figure 9.4 System diagram for general FIR filter implementation.

$$y(n) = g \cdot \sum_{k=0}^{N-1} x(n-k) \cdot a(k) \tag{9.17}$$

9.3.1 Real-Time Implementation of FIR Filters

Although the algorithm would appear to be very simple for such an implementation, the fact that we require $N - 1$ values to store the memory states (past values of the input) of the filter complicates the procedure somewhat. We will begin by discussing the real-time implementation problem, and then later we will see that significant time savings can be made with nonreal-time implementations. First, we can generate C code to implement Equation 9.17. We will need to update each memory state after each convolutional sum which we can do within the same loop if we perform the convolution in reverse order. The simplest way to handle this reverse order is to reverse the coefficients and memory states in their respective arrays as shown in Equations 9.18 (the coefficient array) and Equation 9.19 (the memory state array). Note that the gain constant is stored as the last entry in the C array.

$$C \text{ array: } \{a_{N-1}, a_{N-2}, \ldots, a_1, a_0, gain \tag{9.18}$$

$$M \text{ array: } \{x_{-(N-1)}, x_{-(N-2)}, \ldots, x_{-1}, x_0 \tag{9.19}$$

The segment of code given in Listing 9.3 shows the basic implementation. We start the process by placing the current value of input `x[0]` into the last position of the memory array *M*. The initial value of the output `o` is calculated using the initial values of the memory state and the coefficient arrays. Then we enter the `for` loop to compute the rest of the convolution sum. Each time through the loop, the memory states are updated so that by the end of the loop all of the memory states occupy the correct position for the next input value. The final step in the process is to multiply the output value by the gain constant of the filter.

```
m[N-1]  = x;
o = a[0] * m[0];
for(k = 1; k <= N ;k++)
{ m[k-1] = m[k];
  o += a[k] * m[k];
}
o *= gain;
```

Listing 9.3 Segment of code for FIR filter implementation.

The code of Listing 9.3 will correctly compute the output value but the execution will be slow because of the need for so many index calculations. Listing 9.4 shows the Dig_FIR_Filt_RT function for implementing a real-time (RT) digital FIR filter which makes use of pointers to speed the process. In this case, we will need an `M` array of a size equal to the length of the filter (*N*) and a `C` array which is one larger ($N + 1$) to accommodate the gain constant. Initially, the `M` array is filled with zeros (unless we know the state of the filter), and the C array is filled with the FIR coefficients in reverse order. The final value of the C array is the filter's gain constant. We will use movable pointers within the M and C arrays. The m1 and m2 pointers are initially set to the start of the array and then incremented toward the end. By using two movable pointers, we can transfer memory states without any need for address computation. The final step in the process (before converting the output variable back to an integer) is to multiply the accumulated value in o by the gain constant of the filter. Although for this code, we obtain the value of input from an array, it could just as well be coming from an input port on a DSP system. Likewise, the output value could be written to an output port instead of an array.

```
/*=========================================================
  Dig_FIR_Filt_RT() - filters input array using FIR
               coefs (uses real-time code)
  Prototype:   void Dig_FIR_Filt_RT(int *X,int *Y,
             double *M,double *C,int numb_coefs,int N);
  Return:      error value.
  Arguments:   X - ptr to input array
               Y - ptr to output array
               M - ptr to memory array
               C - ptr to coefs array
               numb_coefs - number of coefficients
               N - number of values in array
=========================================================*/
void Dig_FIR_Filt_RT(int *X,int *Y,double *M,double *C
                                ,int numb_coefs,int N)
{ int      *x,*y,     /*  ptrs to in/out arrays */
           i,j;       /*  loop counters */
  double   *c,*m1,*m2, /*  ptrs to coef/memory array */
           o;         /*  output value */

  /*  Make copies of input and output pointers */
  x = X;
  y = Y;
  /*  Start loop for number of data values */
  for(i = 0; i < N ;i++)
  { /*  Make copy of pointers and start loop */
    M[numb_coefs-1] = *x++;
    c = C;
    m1 = m2 = M;
    o = *m1++ * *c++;
```

```
        /*  Use convolution method for computation */
        for(j = 1; j < numb_coefs ;j++)
        { *m2++ = *m1;
          o += *m1++ * *c++;
        }
        /*  Multiply by gain, convert to int and store */
        o *= *c;
        *y++ = (int)ceil(o-0.5);
    }
}
```

Listing 9.4 Dig_FIR_Filt_RT function.

This implementation of the FIR filter is slow because of the constant memory state shuffle. It should be noted that most sophisticated DSP processors have handled this shuffle by implementing what is referred to as a circular buffer. A circular buffer is one in which the last entry is magically connected to the first entry. In other words, if a pointer which is pointing to the last entry in the buffer is incremented, the processor automatically adjusts the pointer to point to the first entry of the buffer. This helps tremendously, because with a circular buffer, no memory states need to be moved. The newest input value is simply written into the circular buffer over the oldest memory state. The starting point of the convolution is then adjusted to the proper value and the process continues in the normal manner. Although a circular buffer can be simulated on a PC, it requires special coding which is beyond the scope of this text. In the next section we will discuss a much faster nonreal-time method which we can use instead. We will leave the real-time circular buffer implementation to the DSP chip programmers since DSP systems are usually used for this type of work.

9.3.2 Nonreal-Time Implementation of FIR Filters

The implementation of the FIR filter under nonreal-time conditions can be made much more efficient because we will have available as many of the input values as we would like (and that memory will hold). The input values not only represent the input for the system, but also the memory states of the system. The need for the constant shuffle of past memory states will be removed (to a great degree) as we will see. To understand the efficiency of the nonreal-time process, it may be helpful to view the convolution process graphically. Figure 9.5 shows an array of input values and an array of coefficient values. We can visualize the convolution process as the generation of the sum of products as the coefficients slide past the input values.

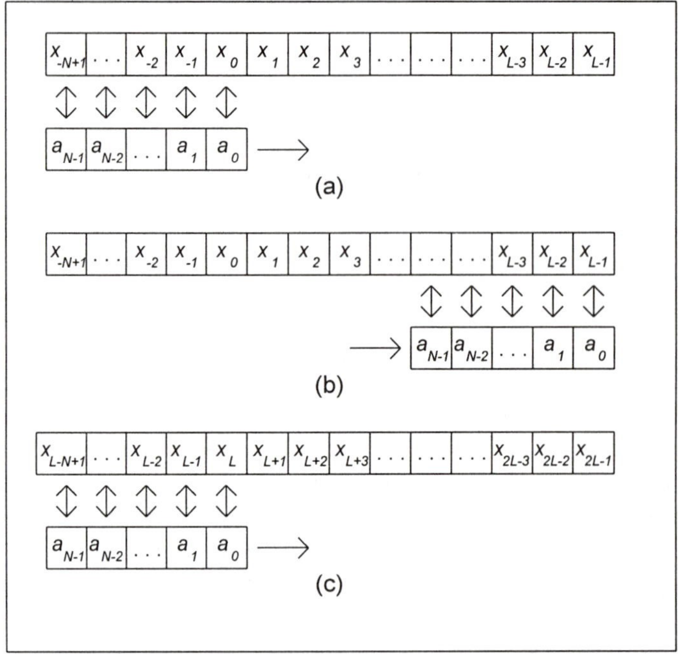

Figure 9.5 Graphical interpretation of convolution.

In part (a) of Figure 9.5, we see the initial position of the convolution process where x_0 is the first input value to be processed. If previous values of the input are not available then x_{-1} through $x_{-(N-1)}$ would be set to zero. After the convolution sum of products is calculated, the value of y_0 can be determined. The array of coefficients can then be effectively moved one position to the right and the convolution performed again for y_1. This procedure will continue until we reach the end of the input array, processing the last value of the array x_{L-1} as shown in part (b). Note that since we have a very large array of input values, we can perform many convolution sums without shuffling the past values of input. However, there is usually a limit to how large an input array can be brought into memory. Therefore, we will eventually need to bring in a new array of L input values and start the convolution again from the start of the new array. But in order for the initial convolution sums on this new array to be correct, the values immediately preceding the x_L value must be available at the beginning of the input array as shown in part (c) of the figure. Consequently, we will need to move the last $N - 1$ values of the initial input array to the beginning on the array. This movement process will have to be accomplished at the end of processing each segment of the input array. If we make the input array buffer much larger than the length of the FIR filter, this amount of processing should be insignificant.

Listing 9.5 shows the `Dig_FIR_Filt_NRT` function for implementing the nonreal-time (NRT) form of the FIR filter. In this function, the pointers to the input, output, and coefficient arrays are passed as arguments as well as the number of coefficients and number of values to be processed in the input array. Notice that a separate array of memory states is not necessary since the input array also represents the memory states. The function starts by assigning the address of the first input value to the pointer variable x which will march through the array keeping track of the starting point of the convolution. A temporary pointer to the proper output array value is also initialized. The pointer to the coefficient array `c` is initialized within the first `for` loop, which controls the processing of all input values. The convolution sum is then calculated moving the x and c pointers progressively through the respective arrays accumulating products. After the summation is complete, the next step is to multiply the accumulation by the filter's gain constant, which has been stored at the last entry in the C array. The process is completed by converting the output floating point value to a fixed point value and resetting the starting point in the x array to the correct position for the calculation of the next output value. After all values in the input array have been processed, the last $N - 1$ values in the array are copied to the beginning of the X array for the processing of the next input segment. As we will see in the next section, when input values are stored in the X array, they are not placed at the beginning of the array (which would overwrite the memory states of the filter), but rather placed at an advanced position in the array based on the length of the filter.

```
/*=========================================================
  Dig_FIR_Filt_NRT() - filters input array using FIR
               coefs (uses nonreal-time code)
  Prototype:   void Dig_FIR_Filt_NRT(int *X,int *Y,
                       double *C,int numb_coefs,int N);
  Return:      error value.
  Arguments:   X - ptr to input array
               Y - ptr to output array
               C - ptr to coefs array
               numb_coefs - number of coefficients
               N - number of values in array
=========================================================*/
```

```
void Dig_FIR_Filt_NRT(int *X,int *Y,double *C,
                                    int numb_coefs,int N)
{ int     *x,*y,    /*  ptrs to in/out arrays */
          i,j;      /*  loop counters */
  double  *c,       /*  ptrs to coefficient array */
          o;        /*  output value */

  /*  Make copies of input and output pointers */
  x = X;
  y = Y;
  /*  Start loop for number of data values */
  for(i = 0; i < N ;i++)
  { /*  Make copy of coefs pointer and start loop */
    c = C;
    o = *x++ * *c++;
    /*  Use convolution method for computation */
    for(j = 1; j < numb_coefs ;j++)
    { o += *x++ * *c++;}
    /*  Multiply by gain, convert to int and store */
    o *= *c;
    *y++ = (int)ceil(o-0.5);
    /*  Reset the pointer in input data */
    x -= (numb_coefs - 1);
  }
  /*  Copy last values to front of buffer */
  memcpy(X,&X[CHUNK_SIZE],numb_coefs-1);
}
```

Listing 9.5 Dig_FIR_Filt_NRT function.

9.4 Filtering and Playing Sound Files

We have now developed several ways to implement the digital filters which we designed in the previous chapters. It is now time to actually implement them and listen to the results. Dedicated DSP processors may not be available to us, but many of us do have sound cards in our computers, so we can at least implement the nonreal-time versions of the filters. The manner in which this will be accomplished is the following. Four sound files have been included on the software disk included with this text, and we can record other sound files using our sound cards. These sound files can be filtered using DIGITAL, the filtering program to be developed in this section.

The filter coefficients we use have been determined using FILTER. After filtering the sound files, we can compare the results using the sound card to play the original and filtered versions of the sound files. (Please read the documentation for your sound card to determine how to record and play sound files.)

9.4.1 Popular Sound File Formats

There are a number of sound file formats in use today, but two of the most popular are the WAVE file format (.WAV) from Microsoft Corp. and the VOICE file format (.VOC) from Creative Labs, Inc. Each of these file formats has a number of different ways that the file information can be stored, but we will concentrate on just the basic techniques. We will discuss only the formats for monaural and stereo signals with either 8-bits or 16-bits per sample. Compression schemes available to either format will not be considered in this text. We will see that handling four different options will provide us with enough challenge for now.

Each of the file formats use the same basic procedure for storing the information necessary to read or write a sound file. Each sound file begins with a header of information which describes the important characteristics of the file such as sampling frequency, number of samples, number of channels (mono or stereo) and number of bits per sample. For our work, the header information for each file can be described as described below.

After the header information, the raw data for the sound file is provided in one of the four formats. If the data file is monaural, the data is just a sequence of bytes or integers depending on the number of bits per sample. The number of data *values* can be determined from the information in the header which specifies the number of data *bytes* in the file. If the file is using 16 bits (2 bytes) per sample, then the number of data values is one-half of the number of data bytes. In the case of a stereo sound file, the byte or integer samples of each channel are alternated starting with the left channel and then the right channel. So if we need to know how many samples to process for the left channel of a 16 bits per sample stereo sound file, we would need to divide the number of data bytes by four to arrive at the proper value.

If the data is in the form of 16 bits per sample, then we can treat the data as a simple signed integer which has a range of values from +32,767 to –32,768. If the data is in the form of 8 bits per sample, then it is stored as an unsigned character with values from 0 to 255 with 128 considered as the midpoint. This is *not* the same as a signed character data type, and therefore

File Format for .WAV Files

Bytes	Description
0 – 3	"RIFF" — identification string
4 – 7	Reserved
8 – 15	"WAVEfmt∅" — identification string (∅ = space)
16 – 19	Reserved
20 – 21	Type of format — short integer
22 – 23	Number of channels — short integer
24 – 27	Number of samples per second — long integer
28 – 31	Average number of bytes per second — long integer
32 – 33	Block alignment — short integer
34 – 35	Number of bits per sample — short integer
36 – 39	"data" — identification string
40 – 43	Number of data bytes — long integer

File Format for .VOC Files

Bytes	Description
0 – 19	"Creative Voice File⊗" — identification string (⊗ = $1A_{16}$)
20 – 21	Offset of data block — short integer
22 – 23	Format version— short integer
24 – 25	$111F_{16}$ — identification code
26	Type of data block — character
27 – 29	Number of data bytes — three characters
30 – 33	Number of samples per second — long integer
34	Number of bits per sample — character
35	Number of channels — character
36 – 37	Type of format — short integer
38 – 41	Reserved

special consideration must be given to the conversion of 8-bit to 16-bit representation. Equation 9.20 indicates the proper procedure to convert from 8-bit unsigned data to 16-bit signed data, while Equation 9.21 shows the opposite conversion. (The functions `Convert2Char` and `Convert2Int` handle this process in the DIGITAL program.)

$$\text{data}_{16} = (\text{data}_8 - 128) \cdot 256 \tag{9.20}$$

$$\text{data}_8 = (\text{data}_{16} / 256) + 128 \tag{9.21}$$

9.4.2 C Code for Filtering Sound Files

We are now ready to develop a complete program to filter either .WAV or .VOC sound files. As we have in the past, we start by making a project outline to organize the development process. We will need to determine the parameters of the filter as well as the waveform to be filtered. Then we will need to read in the waveform data, filter it, and write out the processed data. Since there are four different formats which could be used, and since we don't want to generate four different filtering algorithms to handle each one, we standardize each file type into a monaural 16-bit data waveform. Then, the filtering algorithms can be optimized to operate on that type of file. Since we will be operating in a nonreal-time mode, this conversion should not be a problem.

Digital Filter Implementation Project

I. Determine filter parameters.
II. Determine waveform parameters.
III. Read in and convert data to monaural 16-bit data, if necessary.
IV. Digitally filter the data.
V. Convert data back to original form, if necessary, and write out.
VI. Repeat from III until done.

As we have found in the past, the initial project outline is sufficient to globally define the project requirements, but further details are necessary. As shown below, each of the major parts of the outline can be further subdivided. For example, in order to determine the filter and waveform parameters, we must determine which files we are using in the process (ask the user to specify), read the parameters and display them for user verification. In the case of reading the input data and writing the output data, we must handle the conversion of 8 bits per sample and stereo files. If the file uses 8 bits per sample, the input data waveform is first converted to 16 bits per sample. If the input file is stereo, we then separate it into two monaural files and process them as two independent files. The filtering process will be relatively easy since we have already developed the functions to accomplish this in earlier sections of this chapter.

The `main` function for DIGITAL is shown in Listing 9.6. Some of the routine code has been omitted to shorten the listing. The entire function, as well as all of the others mentioned in this section, can be found in the \DIGITAL\DIGITAL.C module on the software disk. After displaying the opening screen, the program will ask for the names of the filter coefficient file, and the input and output sound files. The data files are then opened as binary read or write files as appropriate.

Digital Filter Implementation Project

I. Determine filter parameters.
   A. Determine which file contains filter parameters.
   B. Read the parameters from the file.
   C. Display the filter parameters for user verification.
II. Determine waveform parameters.
   A. Determine which file contains waveform parameters.
   B. Read the parameters from the waveform.
   C. Display the waveform parameters for user verification.
III. Read in and convert data to monaural 16-bit data, if necessary.
   A. Read in chunk of data to memory buffer.
   B. If waveform uses 8 bits per sample,
      convert to 16 bits per sample.
   C. If waveform is stereo, separate into two monaural waveforms.
IV. Digitally filter the data.
   A. Determine if specified filter is FIR or IIR.
   B. Filter waveform(s) using appropriate algorithm.
V. Convert data back to original form, if necessary, and write out.
   A. If waveform was stereo, convert back from two monaural.
   B. If waveform was 8 bits per sample,
      convert from 16 bits per sample.
   C. Write out chunk of data from memory buffer.
VI. Repeat from III until done.

```
/*=========================================================
  MAIN()
=========================================================*/
void main(void)
{ /* Declare and initialize variables */
  { code omitted }
  /*  Display the opening screen */
  { code omitted }
  while(1)
  { /*  Get filenames from user */
    { code omitted }
    /*  Alloc memory for the structures */
    { code omitted }
    /*  Read filter parameters */
    Error = Read_Filter_Params(FP,descript,CoefFile);
```

```
        if(Error)
        { Clean_Up(filename,FP,WP);
          printf("\n\tFile not found or incorrect type!");
          Bail_Out("Error in Read_Filter_Params!",Error);
        }
        /*  Read waveform parameters */
        Error = Read_Wvfrm_Params(WP,InFile);
        if(Error)
        { Clean_Up(filename,FP,WP);
          printf("\n\tFile not found or incorrect type!");
          Bail_Out("Error in Read_Wvfrm_Params!",Error);
        }
        /*  Write the waveform params to output file */
        Error = Write_Wvfrm_Params(WP,OutFile);
        if(Error)
        { Clean_Up(filename,FP,WP);
          Bail_Out("Error in Write_Wvfrm_Params!",Error);
        }
        /*  Display filter and waveform parameters */
        { code omitted }
        Pause();
        /*  Filter the signal digitally */
        printf("\nWaveform is being filtered: ");
        Error = Digital_Filter(InFile,OutFile,FP,WP);
        if(Error)
        { Clean_Up(filename,FP,WP);
          printf("\n\tFile type may be incorrect! ");
          Bail_Out("Error in Digital_Filter!",Error);
        }
        printf("\b Process complete!\a");
        /*  Free struct memory and close files */
        Clean_Up(filename,FP,WP);
        /*  Run program again? */
        ans = Get_YN(
          "\n\n\t\t    Filter another waveform? (Y/N): ");
        if(ans == 'N') { break;}
    }
    Clear_Screen();
    Cursor_Home();
    Bail_Out("",0);
}
```

Listing 9.6 DIGITAL.C main function.

Next, memory is allocated for a Filt_Params structure to hold the filter information as well as a Wvfrm_Params structure to hold the waveform information. The Wvfrm_Params structure, as defined in DIGITAL.H, contains all of the important information about the waveform.

```
typedef struct
{ char  *header;              /* ptr to header info */
  int   file_type,            /* type of data file */
        numb_chan,            /* number of channels */
        bytes_per_samp;       /* bytes per sample */
  long  samp_per_sec,         /* samples per second */
        numb_samples;         /* number of samples */
} Wvfrm_Params;
```

Listing 9.7 Wvfrm_Params structure.

After the memory has been allocated for the structures, the filter parameters and the waveform parameters are read and displayed. The display of the parameters gives the user a chance to verify the characteristics of the filter and waveform before actually performing the filtering. The display of the parameters could be omitted if desired.

At this point we are ready to perform the digital filtering by calling Digital_Filter, which takes as arguments pointers to the input and output files as well as pointers to the filter and waveform structures. Once the function has completed its work, a completion message is displayed, the Clean_Up function is called, and the user is asked if another file should be filtered. The Clean_Up function, which is used through the main function in case of error, frees all of the memory which has been allocated in the main function, and closes all open files.

The Digital_Filter function performs most of the work in the DIGITAL program and is quite long. Therefore, it will not be listed here but is available on the software disk and can be viewed using a text editor or word processor. In the function, X1 and X2 will be arrays for integer input values, Y1 and Y2 will be arrays for integer output values, and Z will be an array for character values. The coefficient and memory state arrays will be referred to as C and M. The sizes of the data arrays are controlled by CHUNK_SIZE which is defined in DIGITAL.H. The sizes of the X arrays differ based on FIR and IIR filtering because of the addition of room for past input values at the beginning of the array for the FIR case.

After the memory has been allocated for the various arrays, the actual work of filtering begins when the process enters the while loop. It will remain in the loop until the end of the input file has been detected. As indicated earlier, if the input data is in the 8 bits per sample form, the `Convert2Int` function will be used to convert the data to 16 bits per sample form. Then if the input is in stereo form, the `Convert2Mono` function will be used to separate the stereo array into two separate mono arrays.

The actual digital filtering is handled next by using one of the functions discussed in earlier sections. Based on the filter implementation type and the status of the `REAL_TIME` constant (defined in DIGITAL.H), we will use `Dig_IIR_Filter`, `Dig_FIR_Filt_RT`, or `Dig_FIR_Filt_NRT` to produce the filtering. If the input data was stereo, then the respective filtering function will be called twice, once for each channel of information. After the current segment of input data has been filtered, the output data is converted back to the proper form and written to the output file. Following the writing, the loop repeats its process for the next segment of input data. Once the input file has been completely filtered, the allocated memory is freed and control returns to the main function.

9.4.3 Compiling and Running DIGITAL

All of the files necessary to generate the DIGITAL executable are on the software disk included with this text in the \DIGITAL directory. These include the DIGITAL.C, GET_INFO.C, and UTILITY.C modules containing the source code and the associated header (*.H) files. Either a make file (*.MAK) or a project file (*.PRJ) should be prepared with the indicated source files. A **large** memory model should be used to allow for large arrays of data, but no other special requirements are necessary. A copy of the DIGITAL executable is provided in the root directory of the software disk if the user does not wish to compile and link the indicated files.

In addition, as discussed in Chapter 6, there are four sound files included in the \SOUND directory of the software disk accompanying this text. These files represent two different recordings, each provided in the .WAV and .VOC file format. Any sound card should be able to either play one or both of these formats, or have a conversion utility to convert these files to a format which can be played. If file conversion is necessary, wait to convert the files until after the files have been filtered in the examples below. The following demonstrations will require the use of a sound card in order to hear the effects of the filters we are about to design.

In our first demonstration, we will design Butterworth IIR and Parks-McClellan FIR lowpass and highpass filters to satisfy the following characteristics:

LP: $a_{pass} = -0.5$ dB, $a_{stop} = -60$ dB, $f_{pass} = 800$ Hz, and $f_{stop} = 1600$ Hz
HP: $a_{pass} = -0.5$ dB, $a_{stop} = -60$ dB, $f_{pass} = 800$ Hz, and $f_{stop} = 400$ Hz

We will then use these filters to process the MUSIC sound file which uses a sampling frequency of 22,050 Hz. (Either the .WAV or the .VOC format can be used.) After the waveform has been filtered, we can compare the responses in terms of processing time and listening quality. We must use FILTER to design the filters indicated before we can proceed. The coefficients and other data for the Butterworth and Parks-McClellan filters can be saved in the files called LP_BUTR, HP_BUTR, LP_PKMC and HP_PKMC. (The correct extensions will be added by FILTER.)

The magnitude and phase responses of the various filters can be displayed by FILTER. Although not shown here, both the lowpass and highpass Butterworth IIR filters require 12th-order transfer functions which provide –0.5 dB gain at the passband edge frequencies. The lowpass filter provides –64.5 dB at 1600 Hz, while the highpass filter gives –63.5 dB at 400 Hz. The Parks-McClellan FIR lowpass filter requires a length of 63 (up from the predicted value of 57) to provide –0.45 dB and –60.86 dB at the edge frequencies. The highpass filter length prediction was accurate at a length of 111. The gains at the edge frequencies were –0.47 dB and –60.56 dB.

After the filters have been designed, we can use DIGITAL to filter the MUSIC file using the respective .DTB data files for each filter. Figure 9.6 shows the input screen for DIGITAL where filenames are solicited from the user. The output waveform files for this filtering process can be called MUSC_LPB.WAV, MUSC_HPB.WAV, MUSC_LPP.WAV, and MUSC_HPP.WAV, where the last three characters indicate the filter which was used for the processing.

```
******************* DIGITAL - Implementation Program ********************
********************* Copyright (c) 1994  Les Thede **********************


Enter the filename (without extension)
   of a data file generated by FILTER: LP_BUTR.DTB

WAV OR VOC files can be processed.
Enter input filename with extension:   MUSIC.WAV
Enter output filename with extension:  MUSC_LPB.WAV
```

Figure 9.6 Input screen for DIGITAL.

Figure 9.7 shows the display of the filter and waveform parameters as well as the status message which is displayed at the completion of the process. (A key must be pressed to start the filtering process after the parameters are displayed.)

```
****************** DIGITAL - Implementation  Program ******************
******************** Copyright (c) 1994  Les Thede ********************

                    Lowpass Butterworth IIR Filter

          Filter implementation is:    IIR (digital)
          Filter approximation is:     Butterworth
          Filter selectivity is:       Lowpass
          Passband gain        (dB):   -0.50
          Stopband gain        (dB):   -60.00
          Passband frequency   (Hz):   800.00
          Stopband frequency   (Hz):   1600.00
          Sampling Frequency (Hz):     22050.00
          Filter Length or Order:      12

          Waveform File Type:          WAV
          Waveform Channels:           1
          Waveform Bits/Sample:        16
          Wvfrm Sampling Freq.:        22050
          Waveform Total Samples:      165024

          Waveform is being filtered:  Process complete!

          Filter another waveform? (Y/N): N
```

Figure 9.7 Parameter screen for DIGITAL.

After all of the filters have been applied to the MUSIC sound file, we can begin the comparison. First, in terms of processing time, we must remember that all computers are different, so only the ratios of the processing times are provided here. Using the IIR filters as our base, the FIR lowpass nonreal-time implementation (length 65) took about 25% longer to process, while the highpass nonreal-time processing (length 111) took about twice as long as the IIR 12th-order filters. The FIR real-time implementation (without the aid of a circular buffer) took over three times as long as their nonreal-time counterparts. (When comparing these implementations on your own computer, remember that the MUSIC sound file will take about 7.5 seconds to play on its own.)

When we play the original and processed files, we can notice the success in separating the low frequencies from the high frequencies with either the

IIR or FIR filters. (Notice the absence of the chimes in the lowpass version, and the lack of body in the highpass version.) It's difficult to hear any difference between the two highpass versions, but the lowpass Butterworth IIR seems to be able to attenuate high frequencies a bit better than the Parks-McClellan FIR. This can be attributed to the fact that the IIR provides approximately 4.5 dB more attenuation at the edge frequency and that the roll-off continues as the frequency increases. The PM FIR filter, as we remember, has constant ripple in the stopband, so the attenuation does not monotonically decrease in the stopband.

As another demonstration of digital filter implementation, we can design an elliptic IIR and a Kaiser FIR bandpass filter to satisfy the characteristics shown below. We will use these filters to process the SPEECH sound file which uses a sampling frequency of 11,025 Hz. Again, we can use the FILTER program to design the filters required. The coefficients and other data for the elliptic and Kaiser filters can be saved in the files called BP_ELLP and BP_KAIS.

BP: $a_{\text{pass}} = -1$ dB, $a_{\text{stop1}} = a_{\text{stop1}} = -40$ dB, $f_{\text{pass1}} = 300$ Hz, $f_{\text{pass2}} = 3000$ Hz, $f_{\text{stop1}} = 150$ Hz, and $f_{\text{stop2}} = 5500$ Hz

In the elliptic IIR case, an eighth-order filter was designed and provided –1 dB and –54.3 dB at the edge frequencies. In the case of the Kaiser filter, the attenuation was negligible in the passband and –40.99 and –55.6 dB at the lower and upper stopband frequencies, respectively, for a filter length of 167. Although these values apparently meet the specifications, one of the lower side lobes of the Kaiser response violates the stopband gain requirement. Therefore, the filter was redesigned for a length of 175, where all lobes met the specifications. (This points out the importance of checking the complete frequency response when designing filters.) As we would expect, the processing time of this FIR filter took considerably longer (approximately 10 times longer) than the IIR filter.

When comparing the filtered versions of SPEECH to the original versions, we get a clear understanding of the effects of a telephone network on the human voice. (The bandpass filter's characteristics are approximately those of the standard telephone network.) If we listen carefully to the SPEECH waveform, we can hear background noise even in the original. This is primarily quantization noise due to the use of only 8 bits per sample in the quantization. (The original signal was recorded using 16 bits per sample and very little background noise was present.) When comparing the elliptic IIR filter to the Kaiser FIR filter, we find that the Kaiser filter actually sounds a bit better with more low-frequency response. This is undoubtedly due to the major low-frequency lobe which barely provides 40 dB of attenuation. In comparison, the elliptic filter provides at least 54 dB attenuation at all frequencies.

Now that a couple of examples have been given, the process of digital filter design and implementation can be investigated using any filter we design and any waveform we create or record. Since we now know the characteristics of the .WAV and .VOC waveforms, we can create our own special signals of a technical nature, or just for fun.

9.5 CONCLUSION

We have reached the end of this chapter where the implementation of FIR and IIR filters has been developed. We have developed a fully functional program to filter .WAV or .VOC files, which can be mono or stereo, and 8-bit or 16-bit files. Other file formats can be added to the program with a minimum of effort by writing functions to read and write the file headers and passing the required information to our functions. Compressed files can be handled by implementing a decompression function between the reading of the data and its filtering.

Appendix A

Technical References

This appendix includes a list of references appropriate for the study of analog and digital filter design methods using C. Those references marked with an asterisk (*) are particularly useful.

Advanced Mathematics References

*Abramowitz, Milton and Stegun, Irene, eds., *Handbook of Mathematical Functions*, Dover Publications, Inc., New York, 1965.

Cheney, E., *Intro. to Approximation Theory*, McGraw-Hill Book Co., New York, 1966.

Morris, John L., *Computational Methods in Elementary Numerical Analysis*, John Wiley & Sons, New York, 1983.

Rice, John R., *Numerical Methods, Software, and Analysis*, 2nd ed., Academic Press, Inc., San Diego, CA, 1993.

*Spiegel, Murray R., *Mathematical Handbook of Formulas and Tables*, Schaum's Outline Series in Mathematics, McGraw-Hill Book Co., New York, 1968.

Analog Filter Design References

*Daniels, Richard W., *Approximation Methods for Electronic Filter Design*, McGraw-Hill Book Company, New York, 1974.

*Daryanani, Gobind, *Principles of Active Network Synthesis and Design*, John Wiley & Sons, New York, 1976.

Graeme, J. G., Tobey, G. E., and Huelsman, L. P., *Operation Amplifiers*, McGraw-Hill Book Co., New York, 1971

Ghausi, M. S. and Laker, Kenneth R., *Modern Filter Design: Active RC and Switched Capacitor*, Prentice Hall, Inc., Englewood Cliffs, NJ, 1981.

Huelsman, Lawrence P., *Theory and Design of Active RC Circuits*, McGraw-Hill Book Co., New York, 1968.

Johnson, David E., *Introduction to Filter Theory*, Prentice Hall, Inc., Englewood Cliffs, NJ, 1976.

Johnson, D. E., Johnson, J. R., and Moore, H. P., *A Handbook of Active Filters*, Prentice Hall, Inc., Englewood Cliffs, NJ, 1980.

Johnson, Johnny R., *Introduction to Digital Signal Processing*, Prentice Hall, Inc., Englewood Cliffs, NJ, 1989.

Moschytz, G. S. and Horn, P., *Active Filter Design Handbook*, John Wiley & Sons, New York, 1981.

Sedra, Adel S. and Brackett, Peter O., *Filter Theory and Design: Active and Passive*, Matrix Publishers, Inc., Champaign, IL, 1978.

*Van Valkenburg, M. E., *Analog Filter Design*, Holt, Rinehart and Winston, Inc., Chicago, 1982.

Zverev, Anatol, *Handbook of Filter Synthesis*, John Wiley & Sons, New York, 1967.

C Programming References

Bronson, Gary, *C for Engineers and Scientists,* West Publishing Co., New York, 1993.

*Deitel, H. M. and Deitel, P. J., *C How to Program,* 2nd ed., Prentice Hall, Inc., Englewood Cliffs, NJ, 1994.

Hanly, J. R., Koffman, E., and Friedman, F., *Problem Solving and Program Design in C,* Addison-Wesley Publishing Co., Reading, MA, 1993.

*Kernighan, Brian W. and Ritchie, Dennis M., *The C Programming Language,* Prentice Hall, Inc., Englewood Cliffs, NJ, 1978.

*Maguire, Steve, *Writing Solid Code,* Microsoft Press, Redmond, WA, 1993.

*McConnell, Steve, *Code Complete,* Microsoft Press, Redmond, WA, 1993.

Miller, Lawrence H. and Quilici, Alexander E., *Joy of C,* John Wiley & Sons, New York, 1993.

Waite, Mitchell and Prata, Stephen, *The Waite Group's New C Primer Plus,* Howard W. Sams & Co., Carmel, IN, 1990.

Digital Filter Design References

*Antoniou, Andreas, *Digital Filters: Analysis, Design and Applications,* 2nd ed., McGraw-Hill, Inc., New York, 1993.

Bellanger, Maurice, *Digital Processing of Signals: Theory and Practice,* 2nd ed., John Wiley & Sons, New York, 1989.

*Embree, Paul M. and Kimble, Bruce, *C Language Algorithms for Digital Signal Processing,* Prentice Hall, Inc., Englewood Cliffs, NJ, 1991.

*Oppenheim, Alan V. and Schafer, Ronald W., *Digital Signal Processing,* Prentice Hall, Inc., Englewood Cliffs, NJ, 1975.

Oppenheim, Alan V. and Schafer, Ronald W., *Discrete-Time Signal Processing,* Prentice Hall, Inc., Englewood Cliffs, NJ, 1989.

*Parks, T. W. and Burrus, C. S., *Digital Filter Design,* John Wiley & Sons, New York, 1987.

Proakis, John G. and Manolakis, Dimitris G., *Digital Signal Processing: Principles, Algorithms, and Applications,* 2nd ed., Macmillan Publishing Co., New York, 1992.

Programs for Digital Signal Processing, Edited by Digital Signal Processing Committee of IEEE ASSP Society, IEEE Press, New York, 1979.

Roberts, Richard A. and Mullis, Clifford T., *Digital Signal Processing,* Addison-Wesley Publishing Co., Reading, MA, 1987.

Stearns, Samuel D. and David, Ruth A., *Signal Processing Algorithms,* Prentice Hall, Inc., Englewood Cliffs, NJ, 1988.

Strum, Robert D. and Kirk, Donald E., *First Principles of Discrete Systems and Digital Signal Processing,* Addison-Wesley Publishing Co., Reading, MA, 1989.

Appendix B

C Code Organization

This appendix includes a summary of the contents of the C modules on the software disk included with this text. The prototypes for the functions included in each module are provided.

ADV_MATH.C

This module includes functions to compute advanced mathematical functions not found in the standard library. The ADV_MATH.H header file must be available for successful compilation.

```
double acosh(double);
double arcsc(double,double);
double asinh(double);
void   Ellip_Funcs(double,double,double *,double *,double *);
double Ellip_Integral(double);
double Io(double); acosh()
```

ANALOG.C

This module includes functions to implement analog filters in the form of active filters. The ANALOG.H header file must be available for successful compilation.

```
int    Calc_Components(Filt_Params *,RC_Comps *,double,double);
int    Calc_BS_Comps(Filt_Params *,RC_Comps *,double,double);
int    Calc_BP_Comps(Filt_Params *,RC_Comps *,double,double);
int    Calc_LP_Comps(Filt_Params *,RC_Comps *,double,double);
int    Calc_HP_Comps(Filt_Params *,RC_Comps *,double,double);
void   Clean_Up(char *,Filt_Params *,RC_Comps *);
void   Display_Components(RC_Comps *,char *);
int    Display_Filt_Params(Filt_Params *,char *);
void   Display_Header(void);
void   Display_Opening_Screen(void);
char * Get_ANALOG_Basename(void);
int    Read_Filter_Params(Filt_Params *,char *,char *);
int    Write_Circ_File(
           Filt_Params *,RC_Comps *, char *,char *,double,double);
void   Write_BP_Section(int,RC_Comps *,FILE *);
void   Write_BS_Section(int,RC_Comps *,FILE *);
void   Write_HP_Section(int,RC_Comps *,FILE *);
void   Write_LP_Section(int,RC_Comps *,FILE *);
```

B_GRAPHX.C

This module includes functions to implement graphics functions for a Borland compiler. The X_GRAPHX.H header file must be available for successful compilation.

```
void    ClrScrn(void);
int     GetTextHeight(char *);
int     GetTextWidth(char *);
void    LineTo(int,int);
void    MoveTo(int,int);
void    OutText(char *);
void    Rectangle(int,int,int,int);
int     RegisterFont(char *);
void    SetColor(int);
void    SetFontChar(int,int);
void    SetLineStyle(int);
void    SetTextMode(void);
int     SetVGA16Color(void);
void    UnRegisterFont(void);
```

COMPLEX.C

This module includes functions to compute common mathematical operations with complex numbers. The COMPLEX.H header file must be available for successful compilation.

```
complex   cadd(complex,complex);
double    cang(complex);
complex   cconj(complex);
complex   cdiv(complex,complex);
double    cimag(complex);
double    cmag(complex);
complex   cmplx(double,double);
complex   cmul(complex,complex);
complex   cneg(complex);
void      cprt(complex);
void      cQuadratic(complex,complex,complex,complex *,complex *);
double    creal(complex);
complex   csqr(complex);
complex   csub(complex,complex);
```

DIGITAL.C

This module includes functions to implement digital filters (FIR or IIR) on digital computers. The DIGITAL.H header file must be available for successful compilation.

```
void    Clean_Up(Filt_Params *,Wvfrm_Params *);
void    Convert2Char(int *,unsigned char *,int);
void    Convert2Int(unsigned char *,int *,int);
void    Convert2Mono(int *,int *,int);
void    Convert2Stereo(int *,int *,int);
int     Digital_Filter(FILE *,FILE *,Filt_Params *,Wvfrm_Params *);
void    Dig_FIR_Filt_NRT(int *,int *,double *,int,int);
void    Dig_FIR_Filt_RT(int *,int *,double *,double *,int,int);
void    Dig_IIR_Filter(int *,int *,double *,double *,int,int);
int     Display_Filt_Params(Filt_Params *,char *);
void    Display_Header(void);
void    Display_Opening_Screen(void);
void    Display_Wvfrm_Params(Wvfrm_Params *);
char *  Get_DIGITAL_Basename(void);
int     Read_Filter_Params(Filt_Params *,char *,FILE *);
int     Read_VOC_Params(Wvfrm_Params *,FILE *);
int     Read_WAV_Params(Wvfrm_Params *,FILE *);
int     Read_Wvfrm_Params(Wvfrm_Params *,FILE *);
int     Write_Wvfrm_Params(Wvfrm_Params *,FILE *);
```

ERRORNUM.H

This is the header file with defined error values used by many of the C modules. It must be available for successful compilation.

F_DESIGN.C

This module includes functions to design analog and digital filters. The F_DESIGN.H header file must be available for successful compilation.

```
int     Bilinear_Transform(Filt_Params *);
int     Calc_Analog_Coefs(Filt_Params *);
int     Calc_Bart_Win_Coefs(Filt_Params *);
int     Calc_Blck_Win_Coefs(Filt_Params *);
int     Calc_Butter_Coefs(Filt_Params *);
int     Calc_Cheby_Coefs(Filt_Params *);
int     Calc_DigFIR_Coefs(Filt_Params *);
int     Calc_DigIIR_Coefs(Filt_Params *);
```

```
int     Calc_Ellipt_Coefs(Filt_Params *);
int     Calc_Filter_Coefs(Filt_Params *);
int     Calc_Filter_Order(Filt_Params *);
int     Calc_Hamm_Win_Coefs(Filt_Params *);
int     Calc_Hann_Win_Coefs(Filt_Params *);
int     Calc_ICheby_Coefs(Filt_Params *);
int     Calc_Ideal_FIR_Coefs(Filt_Params *);
int     Calc_Kais_Win_Coefs(Filt_Params *,double);
int     Calc_Normal_Coefs(Filt_Params *);
int     Calc_ParkMccl_Coefs(Filt_Params *);
int     Calc_Rect_Win_Coefs(Filt_Params *);
int     Compute_Coefs(Filt_Params *,PkMc_Params *);
void    Compute_Lagrange(PkMc_Params *);
int     Display_Filter_Params(Filt_Params *,char *);
double Estm_Filter_Len(Filt_Params *);
double Estm_Freq_Resp(PkMc_Params *,double);
int     Get_Filter_Specs(Filt_Params *);
int     Mult_Win_Ideal_Coefs(Filt_Params *);
int     Remez_Interchange(PkMc_Params *);
int     Save_Filter_Params(Filt_Params *,char *,char *);
int     Set_PkMc_Params(Filt_Params *,PkMc_Params *);
int     Unnormalize_Coefs(Filt_Params *);
int     Unnorm_BP_Coefs(Filt_Params *,double,double);
int     Unnorm_BS_Coefs(Filt_Params *,double,double);
int     Unnorm_HP_Coefs(Filt_Params *,double);
int     Unnorm_LP_Coefs(Filt_Params *,double);
int     UnWarp_Freqs(Filt_Params *);
int     Warp_Freqs(Filt_Params *);
```

F_RESPON.C

This module includes functions to compute the frequency response of analog and digital filters. The F_RESPON.H header file must be available for successful compilation.

```
int     Calc_Analog_Resp(Filt_Params *,Resp_Params *);
int     Calc_DigFIR_Resp(Filt_Params *,Resp_Params *);
int     Calc_DigIIR_Resp(Filt_Params *,Resp_Params *);
int     Calc_Edge_Resp(Filt_Params *,Resp_Params *);
int     Calc_Filter_Resp(Filt_Params *,Resp_Params *);
void    Display_Edge_Resp(Resp_Params *);
int     Get_Response_Specs(Resp_Params *);
int     Save_Filter_Resp(Resp_Params *,char *,char *);
```

F_SCREEN.C

This module includes functions to display the frequency responses in graphics mode. The F_SCREEN.H header file must be available for successful compilation.

```
int *   Calc_Maj_Pos(plot_area *,char);
int *   Calc_Min_Pos(plot_area *,char);
int     Display_Screen_Data(
            Resp_Params *,Scrn_Params *,char *,double);
void    Limit_Data(plot_data *,double);
int     Load_Labels(Resp_Params *,Scrn_Params *,int);
int     Load_Plot_Area(Resp_Params *,Scrn_Params *,int);
int     Load_Plot_Data(Resp_Params *,Scrn_Params *,int,double);
int     Load_Titles(Resp_Params *,Scrn_Params *,int,char *);
int     Show_Box(plot_area *);
void    Show_Data(Scrn_Params *,int);
int     Show_Div(plot_area *);
int     Show_Label(label *);
int     Show_Text(Scrn_Params *);
int     Show_Title(title *,int);
```

FILTER.C

This module includes the main() function and several general functions for the FILTER design program. The FILTER.H header file must be available for successful compilation.

```
void    Clean_Up(char *,
            char *,Filt_Params *, Resp_Params *,Scrn_Params *,int);
void    Display_Header(void);
void    Display_Opening_Screen(void);
char *  Get_FILTER_Basename(void);
```

GET_INFO.C

This module includes functions to get information from the user via the keyboard. The GET_INFO.H header file must be available for successful compilation.

```
char     Get_Char(char *);
double   Get_Double(char *,double,double);
char *   Get_Fixed_String(char *,int);
int      Get_Int(char *,int,int);
char *   Get_String(char *);
char     Get_YN(char *);
```

M_GRAPHX.C

This module includes functions to implement graphics functions for the Microsoft compiler. M_GRAPHX.H header file must be available for successful compilation.

```
void     ClrScrn(void);
int      GetTextHeight(char *);
int      GetTextWidth(char *);
void     LineTo(int,int);
void     MoveTo(int,int);
void     OutText(char *);
void     Rectangle(int,int,int,int);
int      RegisterFont(char *);
void     SetColor(int);
void     SetFontChar(int,int);
void     SetLineStyle(int);
void     SetTextMode(void);
int      SetVGA16Color(void);
void     UnRegisterFont(void);
```

UTILITY.C

This module includes functions of a general utility nature. The UTILITY.H header file must be available for successful compilation.

```
void     Bail_Out(char *,int);
void     Clear_Screen(void);
void     Cursor_Home(void);
void     Pause(void);
```

X_GRAPHX.C

This module includes functions to implement graphics functions for a PowerC compiler. X_GRAPHX.H header file must be available for successful compilation.

```
void    ClrScrn(void);
int     GetTextHeight(char *);
int     GetTextWidth(char *);
void    LineTo(int,int);
void    MoveTo(int,int);
void    OutText(char *);
void    Rectangle(int,int,int,int);
int     RegisterFont(char *);
void    SetColor(int);
void    SetFontChar(int,int);
void    SetLineStyle(int);
void    SetTextMode(void);
int     SetVGA16Color(void);
void    UnRegisterFont(void);
```

Index

—A—

—B—

—C—

—D—

—N—

—P—

—Q—

—R—

—S—

—T—

—U—

—V—

—W—

—Z—

LICENSE AGREEMENT AND LIMITED WARRANTY

READ THE FOLLOWING TERMS AND CONDITIONS CAREFULLY BEFORE OPENING THIS DISK PACKAGE. THIS LEGAL DOCUMENT IS AN AGREEMENT BETWEEN YOU AND PRENTICE-HALL, INC. (THE "COMPANY"). BY OPENING THIS SEALED DISK PACKAGE, YOU ARE AGREEING TO BE BOUND BY THESE TERMS AND CONDITIONS. IF YOU DO NOT AGREE WITH THESE TERMS AND CONDITIONS, DO NOT OPEN THE DISK PACKAGE. PROMPTLY RETURN THE UNOPENED DISK PACKAGE AND ALL ACCOMPANYING ITEMS TO THE PLACE YOU OBTAINED THEM FOR A FULL REFUND OF ANY SUMS YOU HAVE PAID.

1. **GRANT OF LICENSE:** In consideration of your payment of the license fee, which is part of the price you paid for this product, and your agreement to abide by the terms and conditions of this Agreement, the Company grants to you a nonexclusive right to use and display the copy of the enclosed software program (hereinafter the "SOFTWARE") on a single computer (i.e., with a single CPU) at a single location so long as you comply with the terms of this Agreement. The Company reserves all rights not expressly granted to you under this Agreement.

2. **OWNERSHIP OF SOFTWARE:** You own only the magnetic or physical media (the enclosed disks) on which the SOFTWARE is recorded or fixed, but the Company retains all the rights, title, and ownership to the SOFTWARE recorded on the original disk copy(ies) and all subsequent copies of the SOFTWARE, regardless of the form or media on which the original or other copies may exist. This license is not a sale of the original SOFTWARE or any copy to you.

3. **COPY RESTRICTIONS:** This SOFTWARE and the accompanying printed materials and user manual (the "Documentation") are the subject of copyright. You may not copy the Documentation or the SOFTWARE, except that you may make a single copy of the SOFTWARE for backup or archival purposes only. You may be held legally responsible for any copying or copyright infringement which is caused or encouraged by your failure to abide by the terms of this restriction.

4. **USE RESTRICTIONS:** You may not network the SOFTWARE or otherwise use it on more than one computer or computer terminal at the same time. You may physically transfer the SOFTWARE from one computer to another provided that the SOFTWARE is used on only one computer at a time. You may not distribute copies of the SOFTWARE or Documentation to others. You may not reverse engineer, disassemble, decompile, modify, adapt, translate, or create derivative works based on the SOFTWARE or the Documentation without the prior written consent of the Company.

5. **TRANSFER RESTRICTIONS:** The enclosed SOFTWARE is licensed only to you and may not be transferred to any one else without the prior written consent of the Company. Any unauthorized transfer of the SOFTWARE shall result in the immediate termination of this Agreement.

6. **TERMINATION:** This license is effective until terminated. This license will terminate automatically without notice from the Company and become null and void if you fail to comply with any provisions or limitations of this license. Upon termination, you shall destroy the Documentation and all copies of the SOFTWARE. All provisions of this Agreement as to warranties, limitation of liability, remedies or damages, and our ownership rights shall survive termination.

7. **MISCELLANEOUS:** This Agreement shall be construed in accordance with the laws of the United States of America and the State of New York and shall benefit the Company, its affiliates, and assignees.

8. **LIMITED WARRANTY AND DISCLAIMER OF WARRANTY:** The Company warrants that the SOFTWARE, when properly used in accordance with the Documentation, will operate in substantial

conformity with the description of the SOFTWARE set forth in the Documentation. The Company does not warrant that the SOFTWARE will meet your requirements or that the operation of the SOFTWARE will be uninterrupted or error-free. The Company warrants that the media on which the SOFTWARE is delivered shall be free from defects in materials and workmanship under normal use for a period of thirty (30) days from the date of your purchase. Your only remedy and the Company's only obligation under these limited warranties is, at the Company's option, return of the warranted item for a refund of any amounts paid by you or replacement of the item. Any replacement of SOFTWARE or media under the warranties shall not extend the original warranty period. The limited warranty set forth above shall not apply to any SOFTWARE which the Company determines in good faith has been subject to misuse, neglect, improper installation, repair, alteration, or damage by you. EXCEPT FOR THE EXPRESSED WARRANTIES SET FORTH ABOVE, THE COMPANY DISCLAIMS ALL WARRANTIES, EXPRESS OR IMPLIED, INCLUDING WITHOUT LIMITATION, THE IMPLIED WARRANTIES OF MERCHANTABILITY AND FITNESS FOR A PARTICULAR PURPOSE. EXCEPT FOR THE EXPRESS WARRANTY SET FORTH ABOVE, THE COMPANY DOES NOT WARRANT, GUARANTEE, OR MAKE ANY REPRESENTATION REGARDING THE USE OR THE RESULTS OF THE USE OF THE SOFTWARE IN TERMS OF ITS CORRECTNESS, ACCURACY, RELIABILITY, CURRENTNESS, OR OTHERWISE.

IN NO EVENT, SHALL THE COMPANY OR ITS EMPLOYEES, AGENTS, SUPPLIERS, OR CONTRACTORS BE LIABLE FOR ANY INCIDENTAL, INDIRECT, SPECIAL, OR CONSEQUENTIAL DAMAGES ARISING OUT OF OR IN CONNECTION WITH THE LICENSE GRANTED UNDER THIS AGREEMENT, OR FOR LOSS OF USE, LOSS OF DATA, LOSS OF INCOME OR PROFIT, OR OTHER LOSSES, SUSTAINED AS A RESULT OF INJURY TO ANY PERSON, OR LOSS OF OR DAMAGE TO PROPERTY, OR CLAIMS OF THIRD PARTIES, EVEN IF THE COMPANY OR AN AUTHORIZED REPRESENTATIVE OF THE COMPANY HAS BEEN ADVISED OF THE POSSIBILITY OF SUCH DAMAGES. IN NO EVENT SHALL LIABILITY OF THE COMPANY FOR DAMAGES WITH RESPECT TO THE SOFTWARE EXCEED THE AMOUNTS ACTUALLY PAID BY YOU, IF ANY, FOR THE SOFTWARE.

SOME JURISDICTIONS DO NOT ALLOW THE LIMITATION OF IMPLIED WARRANTIES OR LIABILITY FOR INCIDENTAL, INDIRECT, SPECIAL, OR CONSEQUENTIAL DAMAGES, SO THE ABOVE LIMITATIONS MAY NOT ALWAYS APPLY. THE WARRANTIES IN THIS AGREEMENT GIVE YOU SPECIFIC LEGAL RIGHTS AND YOU MAY ALSO HAVE OTHER RIGHTS WHICH VARY IN ACCORDANCE WITH LOCAL LAW.

ACKNOWLEDGMENT

YOU ACKNOWLEDGE THAT YOU HAVE READ THIS AGREEMENT, UNDERSTAND IT, AND AGREE TO BE BOUND BY ITS TERMS AND CONDITIONS. YOU ALSO AGREE THAT THIS AGREEMENT IS THE COMPLETE AND EXCLUSIVE STATEMENT OF THE AGREEMENT BETWEEN YOU AND THE COMPANY AND SUPERSEDES ALL PROPOSALS OR PRIOR AGREEMENTS, ORAL, OR WRITTEN, AND ANY OTHER COMMUNICATIONS BETWEEN YOU AND THE COMPANY OR ANY REPRESENTATIVE OF THE COMPANY RELATING TO THE SUBJECT MATTER OF THIS AGREEMENT.

Should you have any questions concerning this Agreement or if you wish to contact the Company for any reason, please contact in writing at the address below.

Robin Short
Prentice Hall PTR
One Lake Street
Upper Saddle River, New Jersey 07458